THE
CURIOUS SCIENCE
OF
BODILY FLUIDS

THE
CURIOUS SCIENCE
OF
BODILY FLUIDS

Discover What's Floating
Around Inside of You!

ÅSMUND HUSABØ EIKENES

Translated by Claudia Fox Reppen

Skyhorse Publishing

Copyright © Det Norske Samlaget, Oslo, 2018
Published by Det Norske Samlaget, Oslo
Published with agreement by Hagen Agency, Oslo
English translation © Skyhorse Publishing 2022

Skyhorse Publishing books may be purchased in bulk at special discounts for sales promotion, corporate gifts, fund-raising, or educational purposes. Special editions can also be created to specifications. For details, contact the Special Sales Department, Skyhorse Publishing, 307 West 36th Street, 11th Floor, New York, NY 10018 or info@skyhorsepublishing.com.

Skyhorse® and Skyhorse Publishing® are registered trademarks of Skyhorse Publishing, Inc.®, a Delaware corporation.

Visit our website at www.skyhorsepublishing.com

10 9 8 7 6 5 4 3 2 1

Library of Congress Cataloging-in-Publication Data is available on file.

ISBN: 978-1-5107-5977-0
Ebook ISBN: 978-1-5107-5978-7

Cover design by David Ter-Avanesyan
Cover illustrations: Shutterstock

Printed in the United States of America

CONTENTS

APERITIF

Monday, May 13, 1940. It has been eight months since Germany invaded Poland and Winston Churchill has a formidable task ahead of him: he needs to convince those elected by the people to the House of Commons that he is the right man to assemble the British government in the war that lies ahead. The prime minister promises to do his very best and is willing to give it his all. "I have nothing to offer but blood, toil, tears, and sweat," he says with gravity in his voice, members of Parliament rallying around him.

The quote was later simplified to "blood, sweat, and tears." Churchill had perhaps no more bodily fluids to offer. I do.

In the pages to follow, I have pimple pus, breast milk, semen, and urine on offer, along with bone marrow, spinal fluid, intestinal fluid, and mucous. It will be surprising, and it will be serious. It will be entertaining, informative, and probably a little embarrassing, because there is no escaping the fact that the body can be awkward. Many will say that the body is, above all, something of a mess. They are right. Bodily fluids *are* gross and intimate.

But they are also fascinating and complex.

Fluids are a big part of life, and of the body itself. A person who weighs around 150 pounds contains as much as ten gallons of water. Most of this water is well hidden in the organs and on the inside of each cell, but up to one and a half liters of blood, tissue fluid, spinal fluid, and urine flow freely underneath the skin.

You are perhaps wondering how many bodily fluids are actually in the body. For a short list that is easy to remember, the answer should be between five and ten. There can't really be that many fluids in one body, can there?

It turns out, however, that nature is more complex than you would think. The human body is, after all, the result of evolution over millions of years, so it should not come as a surprise that its solutions are complex. Even though fluids are, to put it very simply, water with some other stuff, it is much more than nuance that separates blood and urine from semen and tears. But the fact that the list includes more than fifty different bodily fluids is both overwhelming and impressive, as well as pretty interesting.

Blood, Sweat, and Tears, and All the Others

One of the reasons why the list is so long is that some of these fluids can be divided into smaller parts. Just as semen can be divided into different fluids made by different glands, so it is with everything that flows in the intestines. Because of their different location, function, or anatomical position, fluids get a different name and number even though they, in certain cases, are very similar to each other.

Complexity makes it so that my list probably includes a few more fluids than you would find in more conservative reference books. And, perhaps, new bodily fluids will eventually appear as we obtain an even more detailed knowledge about our bodies.

For the time being, blood (fluid number 1) is the obvious celebrity among bodily fluids. This is not without reason, as blood both repulses and attracts, symbolically and scientifically.

Blood transports oxygen and nutrition to all the cells, returns with waste, and protects the body against danger. Blood also transports information over long distances, such as messages from the brain or the sex organs that change is occurring. Blood also transports body heat and local energy produced by the heart and internal organs, and it transfers warmth to fingertips and earlobes.

For researchers, blood is not just blood. Blood without red and white blood cells is called plasma (2), a yellow fluid that makes up about half of blood volume. Blood that also lacks coagulation factors is called serum (3). Blood, plasma, and serum are nuances of the same blood, but in detailed reports for educated healthcare personnel, three different bodily fluids are calculated.

All blood is made in the bone marrow (4), a fluid located in our skeletal cavities. Bone marrow is distinct from blood in that it is full of stem cells, bone cells, connective tissue, and fat cells.

A typical body has around half a liter of urine (5) in the bladder and up to about two and a half gallons of tissue fluid, also called lymph (6), dispersed throughout. Tissue fluid surrounds all the cells in the organs and excess fluid is transported back to circulation through the lymphatic system. The lymphatic system is a large network for transport of water, immune cells, and waste; it contributes to the balance of fluid in the body. Tissue fluid keeps the cells healthy and clean and maintains moisture in all the parts of the body.

Every day, two and a half liters of fluid go in and out of the body. Water disappears as sweat (7) and steam with each exhalation; the water must be replaced through food and drink. It starts in the mouth, where resting saliva (8) keeps the mouth clean and fresh. When the jaw begins to chew food, enzymes in the activated saliva (9) help to break down food. Also, in the mouth, nose, and throat, mucous (10) protects surfaces; in the event of an infection, it thickens and fills with dead immune cells, which we call nasal mucous (11). The throat has its own mucous, also called phlegm (12), which helps to keep the respiratory tract free of intruders and particles.

Bodily fluids handle food and drink as if they were on an assembly line. First, gastric acid (13) goes to work in the stomach; if the food comes back up, we call it vomit (14), but for the most part, the food moves on to the small intestine where it is named chyme (15). Here, the pancreas supplements with both a neutralizing fluid (16), which halts the effect of the gastric acid, and bile (17), which breaks down fat; the dissolved fat in the food is called chyle (18) when it has been mixed with lymph. Intestinal juice (19) from the wall of the small intestine makes the mixture of food even more fluid on the way down. Finally, the bowels absorb most of the water, so a normal poop contains only a little fluid. Stool can sometimes be full of water, which we call diarrhea (20).

Many of the over fifty bodily fluids carry out very specialized tasks. The job they do is needed only in one specific place in the body, and we rarely see it. Many would be surprised to know that four fluids in all are hidden on the inside and outside of the eye: the fluid inside the eyeball (21), the fluid between the lens and the cornea (22), the tear fluid that moistens the exterior of the eye (23), and the fluid that hardens into a yellow spot in the corner of the eye every morning (24).

Several of our organs have drops of fluid that protect and lubricate surfaces, collectively called serous fluids. They are not all alike, and usually they are produced by cells in the tissues that require extra protection. Sebum (25) and earwax (26) are both viscous fluids that protect our exterior.

Spinal fluid (27) bathes the brain and spinal cord; a fluid in the ear bathes the sensors of the vestibular system (28), while fluid in the pericardium (29), fluid between the double membrane that surrounds the lungs (30), and fluid in the abdominal cavity (31) each bathe their own organ in the upper body. Joint fluid—also known as synovial fluid—protects the bones and makes a noise when you crack your knuckles (32).

If you injure yourself, small drops of wound fluid (33) will drip out before dead immune cells and a bit of water appear as a yellow-colored pus (34). The skin underneath can also get swollen and painful when fluid from the blood (35) streams to the injured area.

Men have a handful of glands in the reproductive organ that produce fluid. Under the foreskin of the penis are small glands that produce smegma (36), a thick, whitish fluid also nicknamed "dick cheese." Located beside the prostate are two small glands that produce pre-ejaculate (37), small drops that lubricate the head of the penis. Fluid from the prostate (38) and fluid from the seminal vesicles (39) are contributed as important ingredients in semen (40). Newly produced sperm bathe in their own fluid (41) while they wait their turn.

In women, glands in the cervix produce discharge (42), which also increases in amount during intercourse. Some women can also ejaculate (43), though researchers do not all agree on where it comes from, or what it is. Fluid in the fallopian tubes (44) allows the egg to surf down to sperm that are possibly waiting, but if the egg is not fertilized, it is excreted with menstrual blood (45).

In the event that a baby is created in all these fluids, the fetus lies there for nine months, drinking and peeing in the amniotic fluid (46). For the first weeks, the fetus receives help from fluid in the yolk sac (47) to produce blood.

Specialized fluids play an important role for the fetus. A few drops of surfactant (48) in the lungs ensure that the organs develop in the way they are supposed to. On the outside, the skin produces a fatty oil called vernix caseosa (49). Large medical encyclopedias explain that this fluid has a consistency comparable to that of a cheese spread, and that the white cream allows for smooth transport through the birth canal.

The newborn baby drinks the first milk from its mother's breast, called colostrum (50), before being nourished by breast milk for the first few months of its life. On the outside, the first poop is black and sticky, a stool containing swallowed amniotic fluid, dead intestinal cells, and fluid from the bowel, collectively known as meconium (51). For several weeks after giving birth, the mother experiences a discharge known as lochia (52), which contains blood, wound fluid, and mucous membrane remains.

When we die, our bodies become fluid in the end. After cells break down and bacteria consume all they can find of food remains or internal organs, we seep out of the casket as corpse fluid (53) and become a part of nature.

LIFE—BLOODY SERIOUS

If Ingrid Lunde robs a bank, cuts her finger, and leaves behind a few drops of blood, it is not certain that she will be the one who gets the blame. This is because Ingrid's bodily fluids are not like those of other twenty-year-olds.

It starts in the spring of 2016 with a long-lasting throat infection, symptoms resembling those of mononucleosis, and then a sinus infection in her senior year of high school. In August she injures her knee and misses out on her military service as a result, a dream she has had since she was small. At the last minute, she decides to spend a year at a Norwegian folk high school.

Two and a half weeks later, an accident occurs in which Ingrid takes a volleyball at high speed right in the temple. She is dizzy and nauseous as a result. In the emergency room, the doctor concludes she has a concussion and tells her to take it easy for a few days.

Ingrid travels home to Fredrikstad and lies down in her room with the curtains shut, without the strength to do anything other than wait. Two, and then three, weeks pass but the nausea and headache do not. It must be something more than a volleyball and a concussion. At Østfold Hospital they first take some simple blood tests; later they take several more tests that are more extensive.

She is admitted to the hospital, where more of her blood is drawn, as well as bone marrow. The doctor makes a careful conclusion: "You have a serious blood disorder. Most likely cancer."

Ingrid needs new bone marrow.

They Have a Plan

When blood in the arteries isn't doing what it's supposed to, such as in patients with leukemia, a transplant of new bone marrow is a good—though extensive and risky—course of treatment. In the new

1

bone marrow, the patient receives a new blood cell factory—one without dangerous cancer cells.

In the middle of November 2016, Ingrid travels to Oslo University Hospital. All the extra tests have finally come back; the doctors confirm that Ingrid has a special kind of leukemia called acute myeloid leukemia. It is dangerous, but the doctors have a plan.

The oncologist takes time to speak with Ingrid and her parents to explain the program for the coming weeks. He draws up plans for chemotherapy, recovery time, and alternative directions for treatment. They know where to start and have no time to lose.

The same day, Ingrid has an operation to implant a plastic tube known as a central venous catheter, which goes into the right side of her rib cage, an inch under her collarbone. Inside her body, the lines meet her veins, and the catheter makes it possible to send fluid directly into her bloodstream.

For the first seven days, the doctors pump Ingrid's body full of chemotherapy drugs, around the clock, for the entire week. The drugs kill all the cells that are quickly replicating. Many of the cancer cells die, as do the cells that make hair, as well as many of the cells that make mucous membranes in her mouth and intestines.

After a round of chemo, they check her bone marrow for signs of cancer. Ingrid cuts her hair and hopes that it will give her good karma so that she can go home before the next round of chemo. But the test results are not as the doctors hoped: Ingrid is immediately put on round number two of chemo.

This time she develops a host of side effects: the skin on her hands and feet peels off and it is impossible for her to write, to brush her teeth, or to walk. At night she sleeps with ice gloves on her hands. She experiences an outbreak of herpes and is isolated in her little white room with a bed, TV, and bathroom. Christmas comes and goes, a difficult ending to a difficult year.

On the first day of 2017, however, there is good news: her immune system is making a comeback. Seven days later, Ingrid goes home.

Millions of Stem Cells

Ingrid has a younger brother who is a match for being a bone marrow donor, but before she can receive new bone marrow, she needs a third round of chemo. The goal is to destroy her own bone marrow by filling her body full of drugs for six days, followed by a one-day break, and then the transplant.

The same day that Ingrid is lying ready in a room at Oslo University Hospital, her younger brother is lying in a bed at the Radium Hospital. He is under general anesthesia while doctors remove one and a half liters of bone marrow from his hip bone. Three hours later, the bodily fluid is with Ingrid. A nurse connects the bag to the tubes on her chest and millions of stem cells from her younger brother find their way to his sister's skeletal cavities, all on their own.

It takes time for the cells to start making new blood and new immune cells—a risky period for Ingrid. What if her body doesn't accept the foreign cells, even though they came from a young man in her family? "Things could have gone very badly," she comments when we meet six months after her treatment.

While Ingrid waits for an answer about whether her new bone marrow is doing what they hoped, she battles tough aftereffects of the chemo: she throws up every single day for the whole month of February; her appetite is gone; the nausea is back; her lips and cheeks swell up; and she has big canker sores in her mouth. She is in pain and presses gently on the morphine pump.

Doctors and nurses are also waiting for Ingrid's body to react to the new and foreign bone marrow. A reaction to donor tissue is known as a graft-versus-host reaction and it varies from moderate to life-threatening. The doctors want Ingrid to have a mild reaction to the foreign tissue; that way they can control how the new immune system works in tandem with her body.

Ingrid receives an extra dose of immune cells from her younger brother a few months later. A replenishment does the trick, and a period of "seriously squirting diarrhea" indicates that the intestinal immune cells are trying to stabilize. Later, the doctors also think that the immune cells and her lungs are at war with each other, but it turns out to be simply a lung infection. Ingrid has dry mucous membranes and low energy; her immune system is still a work in progress. "Last week I had thrush in my mouth. Constantly, there's one side effect or another from the disease and everything I've been through," she says.

Something More than Before

Ingrid is at high risk for a relapse and the hospital follows up with her almost every week. She has received the best and the most intense treatment for leukemia, profiting from a long series of medical breakthroughs. Without the in-depth knowledge of how blood and bone marrow operate, how cancer forms and threatens patients' lives, and how chemotherapy and a new bodily fluid can be lifesaving, Ingrid would not have had the hope that she does now; even though the long-term effects will have an impact on her daily life going forward, she is both free of cancer and in good health.

She is happy to be active and to lead the normal life of a twenty-year-old: "I have to just live. I can't wait until I get old."

There is also something more to Ingrid now than before she got sick: through her veins flows blood made from her brother's stem cells. They have the same bone marrow, which produces blood with the same genetic material. So, if she robs a bank, cuts her finger, and leaves a few drops of blood behind, it's not certain that she will be the one who gets the blame.

"I think it's kind of cool," Ingrid chuckles. "That it's his bone marrow that kept me alive. It's pretty fascinating."

SPILLAGE—A BODY FULL OF FLUIDS

Two thousand years ago it was believed that there were only four bodily fluids: blood, mucous, and yellow and black bile. Now there are over fifty bodily fluids on the list, a result of comprehensive observations and new discoveries over long periods of time. The path toward the modern view of bodily fluids involved profound theories about life—and a lot of spillage.

In the Beginning, There Were Four

Physicians Hippocrates and Galen believed that four different fluids gave life to human beings. Blood was the source of vitality, and yellow bile in the stomach was necessary to digest food. Mucous was comprised of the clear fluids that cooled and moistened the body, while black bile was only visible when it colored the other fluids, such as in black stool, dark blood, or discoloration of the skin.

Hippocrates is considered the father of modern medicine; he lived from 460 to 377 BC. Galen, the son of an architect, lived from 129 to 216 AD; he believed that medical science, in addition to curing patients, needed to include logical, physical, and ethical thinking. The two left behind hundreds of articles about health, disease, philosophy, and nature, greatly impacting society.

The four fluids had clear parallels to the elements of nature, and the similarities to fire, air, soil, and water explained in elegant fashion how fluids made the body warm, cold, dry, and sweaty. Blood was akin to air and made the body warm and wet, such as during bouts of fever or the sensation of a warm summer wind on the face; yellow bile was warm and dry and resembled fire; mucous was like the cold and wet water, while both soil and black bile were cold and dry.

An ideal body contained a balance of the four fluids, also called cardinal fluids. Since each of the cardinal fluids had its own effect on the body, the variation in fluid balance could also explain the variation between people. Red cheeks or a sweaty brow were the body's way of trying to balance the cardinal fluids.

Body shape was also a result of fluids. Those with an overabundance of mucous were often fat, while those with too much yellow bile were thin and fragile. From here it was a short leap to using cardinal fluids to also explain personal, albeit stereotypical, traits.

The theory of cardinal fluids is also known as the humoral theory, a word with etymological roots demonstrating the cultural influence of the view of the four fluids. The Latin word *humor* stems from humidity or fluid, while the word *temperamentum* depicts a mixture (of liquids) in appropriate proportions (to each other). To be in a particular mood or to have a certain type of temperament are still concepts that characterize who we are or how we behave, even if the biology behind the word is not compatible with modern medicine.

An excess of blood produced an easy-going, cheerful, and bubbly person—a so-called sanguine—while a melancholic person had too much black bile and was gloomy and dreary as a result. Those with too much of the yellow bile—the cholerics—were aggressive and irritable, while those with an excess of mucous—the phlegmatics—were sober, calm, and sometimes rather slow.

The four cardinal fluids provided understandable explanations for everyday bodies, and Hippocrates and Galen expanded the framework to also explain why people got sick. Patients simply had an imbalance in their bodily fluids. Too much blood caused bouts of fever and sweating, while too little blood resulted in a loss of color and energy. An excess of gastric acid, mucous, or black bile caused specific symptoms that doctors could connect to an imbalance in fluids.

Galen suggested that the symptoms would disappear if doctors intervened and corrected the fluid imbalance. Nutritional guidance would ensure that the fluids were kept in check, but sometimes procedures or small operations were necessary. Since an increase in body temperature was the result of an excess of blood, blood-letting was a logical treatment for the symptom of fever. A needle was inserted into a vein in the armpit and blood flowed out of the patient into a tub.

Bloodletting never went out of fashion and many hundreds of years later, the procedure was still the standard treatment for every symptom that could be explained by an excess of warm blood. An adult has around five liters of blood but can survive if he loses about two liters, provided that this does not happen too quickly. From time to time, however, bloodletting would go wrong, and even some well-known people died from the so-called treatment; for instance, the English king, Charles II, died in 1685 of bloodletting as treatment for a sore foot. Napoleon survived having his blood let, but histori-ans believe that a loss of over two liters of blood from bloodletting contributed to the death of President George Washington in 1799.

Blood going astray was much more common in previous times than it is today. However, it would take a long time before anyone understood how blood was transported around the body.

One-Way Journey

Galen was a doctor for the gladiators and got up close to numerous bleeding bodies. The blood that ran from large, open wounds was either red or purple, and the doctor understood that red and purple blood were not the same. He did not understand, however, what this difference meant. Now we know that red, oxygen-rich blood is on its way out into the body via the arteries, while the purple, depleted, and low-oxygen blood is on its way back to the heart via the veins.

To obtain more information about the inside of the body, Galen carried out dissections of apes, sheep, pigs, and goats. His discov-eries from animals were not always correct when he transferred his ideas to people.

Galen thought, among other things, that all blood was produced in the liver. From the liver, the blood set out on a one-way journey through the arteries, was used up, and disappeared out of the body. He believed that the liver's job was to continually make new blood that could supply muscles and organs.

Galen's conclusion was not in line with reality, but no one dared ask questions such as the following: How did the liver, a six-inch-long organ that weighed a pound, manage to produce liters of new blood every single day for seventy years?

The answer would have to wait for hundreds of years. No one was looking or double-checking that the liver was producing new blood at record speed or asking questions about whether there were only four fluids in the body. No one was doing experiments to find out how the blood was getting from the liver out to the arms and legs. For long periods of time, it was monks and priests who filled the role of physician, and because laws and religion limited the opportunities to dissect human corpses, new knowledge about the fluids inside the body was slowed down.

Nearly 1400 years would pass before anyone dared to contradict the views of Hippocrates and Galen on cardinal fluids. When it did finally happen, with the dawning of an interest in observation-based knowledge that characterized the sixteenth century, the discoveries became a snowball of new knowledge that continues to roll today.

The Grave Robber

Andreas Vesalius took drastic measures to get answers to the big questions about what was hiding inside the human body. The Belgian doctor claimed that he stole newly buried bodies and cut down hanged criminals to have something to dissect.

Historians in recent years have moderated the spectacularism of Vesalius's expeditions. Most likely, it was a lot more common to dissect human corpses than how it was portrayed by Vesalius at the time: in the middle of the night he would sneak around local cemeteries, collect body parts, and then carry them back to his laboratory.

Despite the exaggerated depiction of Vesalius as a grave robber, his discoveries are an important milestone in the history of bodily fluids. The new findings were published in 1543 in seven volumes under the title *De humani corporis fabrica* (*On the fabric of the human body*). The books contained detailed descriptions with text and drawings and are an extremely precise and nearly complete reference work about the human body.

Vesalius noticed that both veins and arteries came out of the heart, just as Leonardo da Vinci had seen a hundred years earlier. Their illustrations show a detailed map of where the various components of blood circulation belong. They both understood how central parts of of the puzzle pieced together, but not that the heart was a pump.

The nighttime expeditions of Vesalius were not in vain. At twenty-three he became chief surgeon at the foremost medical school in Europe, just outside of Padova, Italy. Here he built a very important foundation for further discoveries during the sixteenth and seventeenth centuries. Vesalius's goal was to give anatomy instruction a prominent place in the curriculum of a new generation of students, among them William Harvey from Great Britain.

An Islamic Golden Age

In 1597, Harvey moved to Padova to study anatomy. He was a quick learner and saw connections that none of his professors could explain. As early as 1603, Harvey dared to draw the first ground-breaking conclusions about blood circulation based on experiments on live animals: "Movement of the blood occurs constantly in a circular manner and is the result of the beating of the heart," he wrote. The sentence was the beginning of a breakthrough, the start of finally understanding how blood moved around the body.

Blood circulation in us humans is like two laps on a roller coaster, with both laps starting and finishing in the heart. The first loop exits from the right side of the heart, goes through the lungs to pick up oxygen, and then back to the left side of the heart. The next loop exits the left side of the heart, goes out to the whole body via the arteries, through the capillaries into all the tissues, and then back to the right side of the heart via the veins. The blood has then arrived back to the start, ready for two new laps.

Some biologists describe the human heart as two hearts functioning as a single organ. The right side pumps blood out to the lungs and the left side pumps blood out to the body. Easy to understand when one has the answer, but very complex when trying to understand it from dissecting half-rotted animals or buried criminals.

Today, four hundred years after Harvey's breakthrough, it is clear that he was not the first to establish that the heart pumps blood through the lungs, for while Western medicine stood relatively still between Galen and Vesalius, the Arabs made important breakthroughs in the period that some historians refer to as the Islamic Golden Age.

In 1935, a doctoral student discovered several unknown Arabic texts in Berlin State Library. The manuscripts contained, among other things, sharp criticism of Galen's view of anatomy, and were several hundred years older than the texts of Vesalius and Harvey. The author was named Ibn al-Nafis, a Sunni Muslim physician who lived in Cairo from 1213 to 1288. Al-Nafis was just twenty-nine years old when he described the connection between the heart and the lungs.

Al-Nafis understood that the blood in the heart's right side had to make it over to the left side. Galen had suggested that the blood moved from the right side to the left through small, invisible pores in the wall of the heart. Al-Nafis rejected the Western reasoning with the following: "Therefore, the contention of some persons to say that this place is porous is erroneous; it is based on the preconceived idea that the blood from the right ventricle had to pass through this porosity—and they are wrong!"

He wrote instead that blood moved from the right side of the heart out to the lungs where it flowed through small networks in the lung tissue before returning to the left side of the heart. Al-Nafis further argued the presence of small connections between the arteries and veins out in the body during the second loop, an idea that was not confirmed as capillaries until four hundred years later.

Al-Nafis was the first to precisely describe pulmonary circulation, the loop between the heart and the lungs. Harvey reviewed the discoveries and expanded his knowledge with descriptions of how blood got from the heart out to the body, then back again. In the meantime, several Western physicians, including Michael Servetus, Renaldus Columbus, and Juan Valverde, came to same conclusion: the heart pumped blood in two loops, first to the lungs and then out to the rest of the body.

The experiments that formed the basis of this new knowledge were becoming increasingly informative, and to persuade the skeptics, the most ambitious doctors began to use live animals.

The Artery of a Living Dog

Though many contributed, William Harvey holds the status of discoverer of the entire circulation of blood. According to historian Chris Cooper, this is because his texts are written in clear language, and because he used his experiments to argue for his findings. Harvey was an excellent communicator. He was also quite eccentric, illustrated clearly by the story about his experiment on the artery of a living dog.

Dressed in a white coat and hood, Harvey instructed his helpers to tie a living dog down to the dissection table, with its jaw bound to prevent it from barking. Harvey stuck a knife into the dog's rib cage and opened its ribs, displaying the beating heart. As the heart contracted, Harvey cut a main artery. Blood gushed out, spraying the spectators. At the time, it would have been difficult to imagine a clearer way to demonstrate that the heart was a pump.

The historians also tell us that Harvey dissected his own deceased father, and that he later took advantage of the opportunity to expand his knowledge of the opposite sex by dissecting his own sister when she died. His actions strengthen Harvey's image as an objective observer, that which one can perhaps call a "clinically distanced" man.

Harvey published his ideas about blood circulation in the book *Exercitatio Anatomica de Motu Cordis et Sanguinis in Animalibus* (*An Anatomical Exercise on the Motion of the Heart and Blood in Living Beings*) in 1628. With careful calculations, he concluded that the heart pumps around half a liter out from the left side every single minute. Since a person had only about five liters of blood, it was impossible for the body to make all that new blood every ten minutes. Harvey therefore concluded that the blood had to be reused: "It must therefore be concluded that the blood in the animal body moves around in a circle continuously." Based on comprehensive studies and dissections, he explained that the valves of the heart prevented the blood from flowing backward. Blood, therefore, could only move in one direction.

With modern-day measurements it becomes apparent that Harvey was not quite accurate in his calculations. He was sober and cautious in his estimates of size and volume. Careful measurements and the splattering of blood from the dog were enough to convince the skeptics. Now we know that the heart pumps five whole liters of blood out into the body every single minute—much more than Harvey's half liter.

Every minute, almost all the blood in a human body has been through the heart. The blood that leaves the heart has to make it back within a minute, accomplishing this with a speed of about a foot per second. During strenuous exercise, the heart pumps harder and more often. This results in eight gallons of blood per minute streaming out to the body, blood that returns after ten seconds to be sent out from the pump on a new lap at a top speed of almost six and a half feet per second. The total length of all arteries, veins, and capillaries in the body is one hundred thousand kilometers. That is as long as two and a half laps around the equator.

Imagine that. A one-hundred-thousand-kilometer arterial network inside your body.

A Whole New World

Knowledge about the incessant movement of blood inspired new discoveries. At hospitals around Europe, knowledge about anatomy, pathology, chemistry, and pharmacy increased, especially through direct observation of patients and autopsies of the dead. Doctors discovered and charted increasing numbers of fluids and networks, and the internal liquids of the human body played an important role in medical treatments.

In laboratories, doctors carried out experiments with bodily fluids to find out why people got sick. Analysis of bodily fluids gave answers that had been long awaited about how dangerous things, which we now know as bacteria and viruses, spread from one person to another.

In the mid-seventeenth century, Dutchman Anton van Leeuwenhoek examined a few drops under a microscope and discovered small moving clumps, which he called "animalcules." The tiny creatures under Leeuwenhoek's microscope were the start of particle theory, and later knowledge about bacteria and viruses changed the belief that disease was the result of an imbalance of the four cardinal fluids.

Analysis of blood, sweat, tears, and urine also provided us with new knowledge about fertilization, growth, disease, and death.

Doctors began to chart signals in blood of hormones and nutrients that, as if by magic, sent messages for change to all the cells of the body. Eventually, some dared to shift their focus to the intimate parts of the body: the breasts, penis, vagina, and the developing fetus.

This opened up a whole new world.

It became possible to investigate how solid food became liquid nourishment in the intestines, and how the kidneys filtered the blood and produced urine. Brave attempts to transfuse blood from animals to people, and then between people, led to the discovery of blood types, an extremely important breakthrough for medical treatments.

In more recent times, genetic investigations of DNA in every conceivable bodily fluid have contributed to police work and genealogy, and opened up for the controversial technology of gene modification. Bold test subjects pay large sums of money to receive new bodily fluids with the hope of eternal life.

It seems as if we know everything about bodily fluids. Do we?

New Discoveries about Fluids

Even though the modern-day list contains many more—and completely different—fluids than the four cardinal fluids of Hippocrates and Galen, we are nowhere near knowing everything about the liquids of the human body. Take blood, for example: Where does it actually come from?

Recently, American researchers presented a film from the lungs of a mouse where a series of small platelets surprisingly appeared. Platelets are a part of the blood and are important in plugging tears in the arteries. They are produced through a kind of breaking apart of large cells that release parts of their membrane. Up to this point it was believed that it was the bone marrow's job to produce platelets.

Mark Looney and his colleagues from University of California, San Francisco, have used advanced microscopes to film living cells in the lungs of mice. They discovered that the lungs can also produce platelets. The researchers filmed twenty hours of video from ten different mice and counted how many times a large cell produced new platelets. They concluded that each mouse lung produced as many as ten million platelets every single hour, an amount about equal to half of all platelets flowing through the arteries of a mouse at any given time.

Although the results are controversial, Looney is very excited about what the lungs can accomplish: "They are not just for breathing but are also an important partner for making vital components in the blood," he says. Could it be that this applies to our lungs as well? The textbooks tell us, though, that it is the bone marrow that is responsible for the production of platelets. The results from Looney's videos will perhaps necessitate a rewriting and expanding of our knowledge about where blood comes from.

Modern medical research distinguishes itself from research in the old days through the use of advanced instruments, digital analytical methods, and international collaboration. Rapid technological developments give us a much more nuanced and complex picture of the drops inside a human body, and new information constantly pushes the boundaries of truth.

Looney's discovery is based on technology that makes it possible to film fluids and cells on the inside of an organ in a living animal. Other researchers employ methods of moving fluids and cells out of the body and into a laboratory where cells can continue to grow under controlled conditions. In the lab, fluids are easily accessible to curious researchers.

Evatar

By moving the living cells and fluids that surround them into the lab, researchers have full control over everything that happens. At the same time, researchers can avoid ethical challenges related to human experiments. A good example of this is Evatar, an innovation described as "the menstrual cycle on a chip," created by Theresa Woodruff and a large team of researchers from Northwestern University in Chicago.

Evatar is a research tool for exploring reproduction and looks like a LEGO-inspired building set the size of two Kit Kat bars. The building set contains several small, transparent plastic bowls. In them grow cells from mice and humans, cells that live on in the lab as a part of Evatar.

Each plastic bowl functions as a mini version of a central organ in the reproductive system. Cells from an ovary, a fallopian tube, a uterus, and a vagina grow side by side, each in their own plastic bowl, and a blue, artificial bodily fluid flows over the cells and into the plastic tube that connects the organs together.

The blue fluid transports messages from the cells in one plastic bowl to the cells in the next. This collaboration allows Evatar to go through a normal menstrual cycle of twenty-eight days, but since the blood is replaced with an artificial bodily fluid, Evatar does not menstruate.

In the female body, menstruation usually lasts between three and six days. There is large variation between women in how much they bleed, on which days they bleed the most, and how the blood looks. Sometimes it is brown, other times red, and for some it is pink. It can be thin, lumpy, or grainy.

The amount of blood also varies but is on average approximately 1.4 ounces for normal menstruation. Everything between .3 and 2.7 ounces is considered within the normal range, about the size of a chicken egg being a good reference point. An egg is slightly smaller than the uterus, which is about as big as a woman's fist, and the content of an egg is about the same in amount as a normal period.

We also see the blue color of artificial blood in advertisements for pads and tampons. Instead of using a red liquid that resembles blood to demonstrate the absorption qualities of the product, commercials use a blue, sterile, and almost tasteful fluid that disappears into the white material.

The goal of Evatar is not to break down the taboos surrounding menstruation, but rather to contribute to new research. Therefore, in addition to the four reproductive organs, Evatar has a mini liver. The liver is included because its cells can break down the hormone signals that regulate the menstrual cycle. The liver also has a central role in breaking down pharmaceuticals and makes Evatar well suited to investigating how medicines affect each of the organs that are a part of the cycle. By adding a drug to the blue fluid, researchers can test the effect on each mini organ in turn.

Woodruff and her colleagues hope to later connect a mini pancreas and a mini intestine to make their research more complex. Their goal is to make Evatar even more similar to a human body, with a set of carefully selected mini organs that communicate through a blue bodily fluid in the laboratory.

In an interview with the magazine *Wired*, Woodruff states that the technology of cultivating living cells as mini organs in the lab will "radically change the way we study a lot of human systems, not just the female reproductive tract." Now the technology is being used by both cancer researchers and brain scientists.

Brain scientists also use other research methods for exploring the spinal fluid that bathes the brain. The surprising results about fluid in the head have made headlines in recent years.

Water on the Brain

"So many of the simple questions, we don't have answers to. It's very surprising," says researcher Marie Elisabeth Rognes of Simula Research Laboratory in Fornebu, Norway. She has a doctorate in applied mathematics and works with precise descriptions of how water flows through the brain. This is easier said than done because experts are still not in agreement as to where fluid in the brain comes from, or where it's going.

What all brain scientists do agree on is that the brain and spinal cord bathe in spinal fluid, also known as cerebrospinal fluid. The amount is about equal to that of a cup of coffee, and by inserting a needle in the lower back and drawing a few clear drops, doctors can analyze the contents and quality of the liquid to determine whether the patient has a serious illness.

Spinal fluid's most important job is to protect the brain and spinal cord from impact injuries. In addition, spinal fluid nourishes the brain cells and helps to remove waste. Spinal fluid also fills cavities inside the brain, four chambers called ventricles. This is where disagreement arises, explains Rognes.

In recent decades, the majority of scientists have agreed that spinal fluid is produced deep inside one of the ventricles. The hypothesis is that a group of cells called choroid plexus creates spinal fluid by filtering blood from the artery. The choroid plexus pumps out a steady stream of new spinal fluid in the ventricle.

But not everyone holds this traditional view. A smaller group of scientists believe that spinal fluid is made by the brain absorbing a little bit of water from the arteries, an opposing hypothesis to that of a central pump regulating the production of spinal fluid.

The main reason for disagreement is a lack of crystal-clear results from experiments and the various ways of interpreting one's own data, as well as the data of others. That is to say, it is not easy, neither technically nor ethically, to conduct experiments on fluid in the brains of living animals and people. Hypotheses are formed on the basis of many small observations.

This is where Rognes's mathematical models come into the picture. She imagines the brain as an organ governed by physical laws with strict rules for volume, pressure, and movement. In addition to spinal fluid, which surrounds the brain and spinal cord, the brain itself bathes in tissue fluid. In the brain, tissue fluid fills tiny spaces between all the cells, a combined volume that equals nearly one fifth of the entire brain. The two fluids in the head are separated from each other by a membrane that surrounds the brain tissue.

Rognes explains how the brain can be compared to a sponge, where the brain cells are solid material and tissue fluid fills the spaces in between. She and her colleagues study mathematical equations to find out how the fluid moves within the sponge. The goal is to produce precise descriptions of a field of study where all do not agree on how the brain actually is.

An important contribution from Rognes's models is being able to refute new hypotheses about fluid in the brain. "Mathematicians can contribute by putting a number on something," she explains. The numbers related to fluids follow physical laws and are not affected by researchers' personal interests or favorite hypotheses. The goal with numbers is to disprove bad ideas before they steal too much time and energy, and at the same time help researchers get on the right track in their search for answers.

The new results are especially important for learning how to keep the brain healthy, and what goes wrong when people become ill. Research on waste material in the brain is therefore an important topic for Rognes and other brain scientists.

Brainwash

In addition to the fact that brain scientists do not agree on where the fluid comes from, they are also in disagreement when it comes to theories about where the fluid goes when it's "dirty." Dirty fluid, full of cellular waste, can have major consequences.

The brain at work constitutes 20 percent of the body's metabolism. All the activity results in waste that needs to be removed before it damages brain cells. Waste—the remains of metabolism in the brain—must exit the brain. The final stop of this waste is the liver, where it is broken down.

Another type of waste is accumulation of plaque, small clumps of protein that, over many decades, can kill brain cells. Patients with

Alzheimer's disease and Parkinson's disease often have too much plaque in the brain, and many researchers believe that this plaque is an important factor in the disease's progression. Scientists still don't know the details of how the brain cleans itself and washes the plaque away, or what goes wrong when people get sick.

The brain is not like the rest of the body, where the lymphatic system transports an excess of fluid, waste, and immune cells back to the bloodstream. There are no lymphatic arteries deep in the brain, so the organ must carry out the same tasks in another way. The first step is to resolve whether the removal of waste from the brain occurs through passive movement or active transport. Does waste flow out of the head at its own pace, or does the fluid receive help from cells, arteries, and certain transport routes?

An alternative is that waste moves around randomly in tissue fluid. According to some scientists, the dirty fluid flows calmly out of the brain tissue, through the membrane that covers the brain, and into the spinal fluid. Here, the waste flows on, without help, and ends up in a random artery or a nearby lymphatic artery. Finally, it gets to the liver, where it is cut into small pieces. Can it really be as simple as the brain washing itself, without any cleaning help?

In 2013, professor Maiken Nedergaard presented a hypothesis for a system of active cleaning help in the brain where tissue fluid is transported out of the head. Her findings provide new theories on how the brain cleans itself. Nedergaard and her colleagues noticed that tissue fluid in the brain was transported out via tubes surrounding the arteries. Fluid streamed through the tubes from the middle of the brain tissue and out to the surface. The fluid received good help from fluid pumps in a type of nerve cell called glial cells. Since glial cells perform roughly the same task that the lymphatic system does outside of the head, the discovery was named the glymphatic system.

Scientists later discovered that the glymphatic system was most active in sleeping lab animals, and concluded that sleep was important in the removal of dangerous waste material and nerve toxins from the brain. International media published headlines about the discoveries, with simple and elegant descriptions of how the brain gets rid of waste while we sleep.

"Note that this is highly controversial. The glymphatic system has not yet been established as fact," Rognes comments. She sees the developments as a clear example of the cultural differences between medicine and mathematics. It is safe to trust ironclad mathematical evidence, but if someone has observed an interesting result in an experiment with mice, it is not certain that truth will be the same in a few years.

The brainwash becomes even more complicated in that there are lymphatic arteries inside the skulls of both mice and people. In the summer of 2015, two independent research groups concluded that the lymphatic arteries are also important to normal brain function. Both research groups declared their findings as new breakthroughs in anatomical mapping of the brain, but were met with immediate resistance from international colleagues. Ever since Italian Paolo Mascagni mapped out the entire lymphatic system in 1787, lymphatic arteries inside the skull have been known to exist. The lymphatic arteries were later rediscovered and confirmed by other researchers in 1869, 1948, 1953, and 1996. Perhaps it is the lack of electronic PDFs of old research results that leads scientists in 2015 to believe that they are making new discoveries?

The anatomical research showed, however, that the lymphatic arteries cannot take in tissue fluid from the brain itself. There are large distances and several barriers that prevent direct transport from innermost to outermost, as the brain and tissue fluid are well protected by three thin membranes.

Tissue fluid flows in the brain tissue, spinal fluid flows between the two innermost membranes, and the lymphatic arteries lie in between the two outermost membranes. The position of the lymphatic arteries means, though, that they are in the right place at the right time for absorbing some of the spinal fluid. The conclusions are based on mice experiments, and the two research groups are understandably cautious about transferring their results directly to a human brain. Scientists nevertheless speculate about whether the lymphatic arteries outside of the brain are able to participate in cleaning waste from brain cells after the glymphatic system has done its job.

Rognes's approach to the challenges is to delineate the physical laws that govern the flow of fluid and treatment of waste in the brain. She bases her models on data from patients and mice experiments, as well as data from healthy volunteers who lie in the machine that examines fluid flow in the head. Fluid in the head is in constant motion and is affected by the heart beating and by the lungs. "My PhD candidate has perfect fluid flow, while that of some of my colleagues is a bit strange," says Rognes.

When you inhale, the brain and large arteries around the brain expand, and pressure in the head increases. There is not enough room for all the spinal fluid, and a small amount, about as much as a sugar cube, moves down to the spinal cord. When you exhale, the fluid returns to the head. Air in and out, bodily fluid down and up.

Fluid in the head is never at rest. Neither is our knowledge about bodily fluid.

SEX—POPULAR SCIENTIFIC PORN

John and Jane Doe are like most other couples: They have sex one to two times per week. Usually it begins with pressing their lips to each other in a kiss. Maybe they are underneath the covers, or perhaps in the laundry room. Maybe they are on vacation, or perhaps it's a normal weekday. They kiss and forget where they are. She opens her mouth, he sticks out his tongue. In the next ten seconds, John and Jane Doe exchange eighty million bacteria.

A Last Kiss

One summer day in 2012, twenty-one boyfriend–girlfriend couples participated in an attempt to chart how many bacteria move from mouth to mouth. Dutch researcher Remco Kort and his colleagues collected saliva samples from couples before and after a ten-second-long intimate French kiss. The research team used DNA analysis to compare the bacteria flora between the subjects and were not surprised to find that the couples had many of the same bacteria colonies on their tongues.

Afterward, the couples received questions about how often they kiss each other during the course of a day. The women answered that they, on average, had five ten-second kisses, while the men claimed that they kissed their significant other ten times each day. (Researchers reckon that the men exaggerated to boast about their manliness.)

The researchers concluded that couples who have more than nine intimate kisses per day exchange enough bacteria to impact the microbes on their partner's tongue. They need to kiss often because they share bacteria for only a short period of time. Saliva runs down to the stomach and makes it difficult for new bacteria to take root in the mouth or tongue of their better half.

For the next part of the experiment, one person drank a small glass of yogurt before he or she delivered a new saliva sample. Then more kissing, this time with a tongue full of yogurt bacteria. The researchers took a new saliva sample from the person who had not directly tasted the yogurt to count how many of the yogurt bacteria had managed to move from mouth to mouth.

Kort and his colleagues counted the bacteria in the mouth of the recipients before and after the kiss and sorted out the bacteria that had come from the yogurt. Then they did calculations of the tongue's surface where bacteria could attach themselves and of how much yogurt spit had been transferred from the one to the other. The answer was that eighty million bacteria have moved between Jane and John.

The number will vary from person to person, depending on how big your tongue is, how much saliva you produce, and how good you are at tongue gymnastics. Or maybe you think it's safest to go without kissing at all.

There are several theories on why we kiss each other, a field of study called philematology. Lips against lips have cultural and religious interpretations that range from friendship and respect to love and sexual attraction. From an evolutionary perspective, kissing resembles to a certain degree the rituals of birds and mammals half-chewing food for each other and then spitting or throwing up in the mouth of a family member. It sounds more disgusting than it is.

Other theories are based on the idea that kissing allows both the man and the woman to taste a potential partner, thereby checking whether they are compatible to become parents together. Bacteria contribute chemicals that give a taste to saliva. The taste of your boyfriend or girlfriend is actually the taste of bacteria that grow on the tongue and in the mouth of your better half. Maybe you have kissed enough for today.

John and Jane remain unaffected and continue with a coordinated and intense interplay that activates thirty-four different muscles in the face. They tilt their heads carefully to obtain an even deeper kiss. The kiss leads to a number of bodily fluids starting to move. Something is happening.

Pulse increases, body temperature rises, and hormones move around the blood with the signal that good things are about to happen. Both of them feel it between their legs, where increased blood flow makes their genitals expand.

Clitoris—"Knerten"

John Doe's penis is about 4.7 inches long when erect. According to researcher David Veale at King's College in London, John's is slightly under the average of 5.2 inches. In 2014, Veale and his colleagues carried out seventeen different studies with measurements from over fifteen thousand men to find out the normal length and girth of the penis. The conclusion about the average length of an erect penis is based on data from only 692 men. Most of the men were white, and they were not many. The average length probably does not apply to the whole world.

In a book about the male genitalia, *Ett skritt foran* (*One Step Ahead*), physician Volker Wittkamp describes the cross-section of the penis as a sad alien smiley. In the shaft of the penis are three masses of spongey tissue: two that lie beside each other, and one on the underside around the urethra. In a cross-section, it looks like a little frown and two eyes with pupils of blood vessels, according to the German urologist.

Touching the head of the penis and thinking about sex, or just one of the two, leads to the penis becoming erect. Erection is the result of blood that streams to the spongey tissue but doesn't get out again.

Around the blood vessels are thin layers of muscle cells, and in order for the blood vessels to expand, the muscles must relax. Relaxed muscles allow the blood vessels to expand and give room to more blood that makes its way to the spongey tissue. The blood allows the penis to increase in length and girth. But all that extra blood puts pressure on and blocks the veins that normally transport the blood back. It's full stop. John Doe's erect penis is the result of blood that has reached a temporary end station.

In Jane, blood is also streaming to her genitals to fill spongey tissue in the clitoris. Most people know little about how the clitoris looks or functions. The reason is that most of the organ is located inside the body, and because women have received less anatomical attention in a male-dominated field of study. "The main source of female pleasure, the clitoris, is a well-hidden secret, in stark contrast to the, to put it mildly, conspicuously erect penis," wrote Nina Brochmann and Ellen Støkken Dahl in *Gleden med skjeden* (*Pleasure with the Vagina*).

Both males and females share the start of fetal development, and the genitals are based on the same framework. In women, the clitoris is more than the clitoral glans, which Brochmann and Støkken describe as a raisin. The rest of the organ is reminiscent of the *Knerten* figure by Norwegian children's author Anne-Cath. Vestly, with long its arms, rather short stomach, and plump legs. Behind the raisin head, the organ bends itself into an arch, before its arms stretch themselves downward in a slant, and two slightly fatter legs encircle the opening of the vagina. The head, arms, and legs of "Clitoris–*Knerten*" are spongey tissue that increases in size and sensitivity when blood flows into it.

There are no glands in the vagina, so fluid goes through the vaginal wall and into the vagina. The fluid mixes with mucous from the cervix and resembles discharge that cleans the vagina each day, a water-based and usually clear fluid with low pH that flows downward from the cervix. The low pH is detrimental to bacteria and parasites and is an important defense against disease.

In vaginal mucous there lives both yeast and bacteria. They are usually harmless and help the body function normally. If vaginal flora is imbalanced, the symptoms are easily recognizable. For example, wrong types of bacteria produce brown or dark gray discharge that smells like rotten fish, and an undesirable growth of yeast resembles cottage cheese.

One reason why many women can experience an imbalance in the vagina is because of a strong focus on intimate hygiene. Frequent use of intimate soaps and rinsing the vagina with the shower head creates more problems than necessary. "Discharge is the body's own soap," states Marius Johansen, head of medicine at Sex og samfunn (Sex and Society).

The fluid in the vagina is damaging to sperm, intruders that are forced to do their very best to survive. Sperm receive help during the arousal phase from a fluid that drips from the penis, called pre-ejaculate. In some men, it is just a small drop, in others a steady stream.

On the way through the urethra, the pre-ejaculate washes away remains of urine with low pH that can damage the sperm cells. Pre-ejaculate also helps to lubricate the penis so that friction against the vaginal wall is reduced. Even though the drops are not the result of an ejaculation during orgasm, pre-ejaculate can contain a small number of sperm from the previous ejaculation. If the couple has frequent intercourse within a short period of time, sperm in the pre-ejaculate can spoil the strategy of pulling out early.

In the Copy Room

John and Jane have done it many times before, and in their lifetimes, they will have sex between two and three thousand times. How do we know what they do in the bedroom, in the car, at the cottage, or in the copy room?

Because, someone has surveyed the sexual habits of people. Someone took the time to interview thousands of people about sex, a comprehensive study that was quite controversial in the 1940s and 1950s. A central source of knowledge about who, what, and how people have sex is the zoologist Alfred Kinsey.

Together with several colleagues, Kinsey interviewed thousands of men and women about their sex lives. He traveled around to every American state and talked with old, young, married, single, heterosexual, homosexual, and asexual people. In 1948 came a report about the sexual habits of men, based on interviews and stories from twelve thousand men. Five years later came the report about the sexual habits of women, where 5,940 shared their experiences.

The two reports showed a large variation between the interview subjects, and the findings collided with the idea of a standardized and predictable sex life. "There is no sexual behavioral norm that is typically American," wrote editor Albert Deutsch in a summary of the report on women.

Among both men and women, Kinsey and his colleagues found great diversity. There is large variation connected to what gets people aroused, and how often they have sex, either with themselves or with a partner. There is variation in with whom people fall in love, and with whom they have had sexual experience. There is also variation in how sexual habits change over the course of a long life, and what drives men and women in the hunt for a partner. The short version: people are different.

Researchers still can't resist the urge to investigate, measure, survey, and understand what happens in and with a body during sex. Kinsey wanted to take his surveying a step further to observation and study of the sexual act. He dared not apply for grants to conduct such experiments, and neither did he dare to carry them out in his office. However, this did not prevent him from finding an alternative place—his attic at home—where thirty couples volunteered to be filmed. Some of the volunteers also masturbated in front of the camera, including Kinsey himself.

Kinsey was not the first scientist who, for various reasons, participated in, or was accused of being sexually aroused by, research of sexual habits. In the 1890s, gynecologist Robert Latou Dickinson included intimate details about sexual habits in a patient's history before the first consultation. Some of the answers he received were far beyond the norm of what was typical or normal to discuss. Early in the twentieth century, psychologist John Watson moved sex research into the lab. He and a nineteen-year-old student, Rosalie Rayner, who would later become his wife, were perhaps the very first couple to take part as both researchers and test subjects. The audio recording and notes from their experiments allegedly played a central role in the divorce case between Watson and his first wife.

In *Bonk*, Mary Roach explains why so many sex scientists actively participated in their own research projects. The reason was not that they were perverted or abnormal, but rather because it was difficult, embarrassing, and risky to invite strangers to have sex in front of the camera. In the 1940s and 1950s, Kinsey in particular met with strong resistance to his findings. A few decades later, however, conservative attitudes were changing, and scientists took the risk of conducting more and more intimate and detailed experiments.

Imagine this scenario from the early 1960s: A man and a woman sit side by side on a hospital bed. White walls and ceiling lamps light up their naked bodies, where sensors are attached to their heads, breasts, and genitals, wires connected to machines buzzing and beeping. A voice over the speaker gives the signal and the couple begin to touch each other. Behind a one-way mirror are research duo William Masters and Virginia Johnson, who are taking notes. They are eager to confirm a new theory that sexual intercourse has four phases: *arousal, plateau, orgasm, and resolution.*

Mindfulness

John and Jane stopped using contraception six months ago. They want to add a third child to their family, a brother or sister for Kristin (5) and Henry (3). The couple try to have sex on both Tuesdays and Saturdays, unless they are too tired, or if it collides with the days of Jane's menstruation. They know that it usually takes a while before everything falls into place. The couple must be patient and keep trying, even though it varies between the two of them who has the most desire.

"Female desire resides first and foremost in the mind," write Brochmann and Støkken Dahl. They cite a study from 2010, where Meredith Chivers and her colleagues went through 132 research articles before concluding that men and women are different in how the body and mind become aroused. Objective measurements from the penis and clitoris were connected with the participants' subjective answers about how aroused they were.

The result was striking: When men had an erection, they felt, for the most part, that they were aroused. With the women, there was much less consistency between body and mind.

Variation between the test subjects was larger than expected, and the findings are not entirely unambiguous. Experiments affect men and women differently, making it more difficult to draw conclusions. Use of images and video varied between the studies, and there were descriptions of using erotic or explicit material. The test subjects' subjective evaluation for timing of arousal varied between during and afterward, and there were also differences in how many times an individual was allowed to participate in the experiment. Different experiments conclude, of course, that people are different.

Chivers's study results have also previously been controversial. For example, she has reported that while many men get an erection and think of sex when they watch videos with female actors, women do not particularly care about the gender of the actors. Women in one study had little response to the sight of a naked man jogging on the beach, while women in another study had an increase in blood flow to the clitoris by watching apes having sex.

Chivers is now director of the sex and gender laboratory at Queen's University in Ontario, Canada, and continues to study variation and differences in how people have sex. She invites a large selection of people to contribute to her projects by masturbating in a La-Z-Boy chair. In recent years, research articles have been about arousal and sexuality of trans people, as well as people who practice sadomasochism or mindfulness in the comfortable chair. There is still a lot we don't know.

John and Jane, on the other hand, know that practice makes perfect. They also know that being considerate toward each other and communicating during sex makes it better. Jane puts her hand on John's penis and guides it into her in a careful and experienced manner.

The first phase of intercourse, *arousal*, is over, and the couple is on the way to the next phase, *plateau*. The penis and clitoris continue to grow and increase in sensitivity. More blood, more skin against skin, more stimulating movements, feelings, and thoughts. The good friction between them builds up as John thrusts his penis into Jane's vagina, again and again.

They sweat, smile, and kiss each other, passing more and more saliva and bacteria from mouth to mouth, all the while rhythmically working their way closer into each other. With two powerful thrusts, John sends three hundred million sperm cells into Jane's vagina.

Precious Drops

Many have proposed theories about where sperm cells come from. "I assert that seed is secreted from the whole body, from the solid parts and the soft parts, and from all its moisture," wrote Hippocrates over 2300 years ago. Arteries, nerves, tubes, and canals from the entire body were supposedly gathered in the kidneys before semen went through the testicles and out through the penis. Hippocrates was barking up the wrong anatomical tree.

He did not stop there, and Hippocrates's own experience gave him the argument he needed to explain that semen contained the strongest elements of the human body: "When we have sexual relations, even though we release only a small amount of semen, we become tired." When he goes on to explain how the potency of semen determines whether a baby is a boy or a girl, Hippocrates makes his contribution to two thousand years of the oppression of women: "The man is stronger than the woman, and must necessarily be made of the strongest semen." He also makes his contribution to general public health: "By having sexual intercourse with men, [women] will be more healthy. If they do not, they will be less healthy."

In the mid-seventeenth century, it was believed that tiny men lived in sperm cells. The rapid progress of the microscope gave access to a stunning miniature world of blood, tissue, and microorganisms. Some, in their eagerness to find answers to big questions in their microscopic observations, thought that a miniature man in the sperm provided a simple explanation for how the fetus developed inside the woman.

The first one to observe sperm under a microscope was Anton van Leeuwenhoek. "What I investigate is only what [...] remains as a residue after conjugal coitus," he wrote to the academy of science to assure them that he was not doing anything perverse.

If, like van Leeuwenhoek, you have examined fresh semen outside the body, you have perhaps observed that the fluid first coagulates and makes mucousy clumps. After a few minutes, the semen becomes diluted and clear. These changes help the sperm first to get through the cervix, and then swim up into the uterus.

Outside the body, semen acts a bit like egg white. Physician Volker Wittkamp offers therefore the tip that cold water is best for washing away semen from pubic hair (or other hair). If you use hot water, the fluid will coagulate into white and sticky clumps that will only be more difficult to get rid of. Also, there are an awful lot of cells.

An ejaculation contains as many sperm as the entire population of the United States. Together they contain 1500 terabytes of information. In addition to sperm, ejaculate contains a long series of fluid contributions from different parts of the man's reproductive system.

Fluid from the prostate contains nutrients that sperm use, as well as the enzyme that first clumps the semen and then makes it liquid. A couple of pea-sized glands are located right underneath the prostate where they produce a mucousy fluid. Together with the prostate, the pea-sized glands also make fluid that neutralizes the pH of the vagina. The seminal vesicles, two two-inch-long glands located directly underneath the bladder, produce important nutrients for sperm cells, such as fructose and vitamin C. Lastly, a few enzymes also come from the seminal vesicles, which produce prostaglandin to stimulate muscle contractions in the vagina, helping to propel sperm cells forward.

A Gentleman's Marbles

Before getting as far as a uterus, sperm have already experienced a lot. Over two months have passed since production started deep in the testicles. Spherical cells are instructed by testosterone to begin dividing, eventually developing the characteristic form of tadpole-like cells. In each sperm cell, DNA is stored in the head and a motor in the neck controls the movement of the long tail. Nearly fully developed, they gather shoulder to shoulder and push each other onward in a fluid produced by the testicles' helper cells.

A testicle is more than just one half of a pair of marbles. On the top and down the backside of each testicle sits the epididymis, a maturation and storing station for sperm. The epididymis contains an extremely crumpled tube in which sixteen feet is twisted down to three-quarters of an inch. Each and every sperm cell produced in the testicles must be transported through the long tube as it matures along the way. Hundreds of millions of sperm wait in line for their turn. They are not yet able to swim on their own but depend on fluid from the epididymis and tiny hairs in the canals to move forward.

Viewed from the outside, the distance between the testicles and the root of the penis is quite short, but when John ejaculates, the sperm take a 15-inch-long detour. Muscle contractions around each of the two spermatic cords push the sperm up into the abdomen and then over, behind, and under the bladder. Here the two spermatic cords meet each other and go into the prostate, where they are connected to the urethra. Fluid from the seminal vesicles, prostate, and pea-sized glands prepare the sperm for their mission before powerful contractions in the urethra propel them to the last stage.

In the 1950s, many doctors claimed that the force of the ejaculation affected fertility. They claimed that successful sperm were shot like bullets up into the uterus, right into a waiting egg. Kinsey, however, was not convinced.

Proceeding scientifically, Kinsey planned to film two thousand men while they masturbated. He managed to find only three hundred, but the result was nevertheless as he had imagined: a big mess. In three-quarters of the men, the sperm was not shot out in large force, but rather dripped straight down onto the nearest surface. The rest of the men shot not much farther than one and a half feet onto a sheet that covered the floor. The record, on the other hand, was close to eight feet.

Something Absolutely Outstanding

John's climax starts Jane's orgasm. She feels tingling throughout her whole body, muscle contractions in her pelvic region, and a warm feeling of satisfaction. Other times, Jane experiences several orgasms in a row, and from time to time she ejaculates. The first few times, Jane and John were equally surprised by the drops from her urinary tract. The mythical female ejaculation, described as a "fountain orgasm" by *Store medisinske leksikon* (*Norwegian*

Medical Encyclopedia), is not urine, as many believe. The ejaculate differs from urine and discharge in both consistency, smell, and taste, although the origin and function of this fluid is not yet fully understood.

This time is over, without multiple or "fountain" orgasms. Jane can feel how John's penis, which just a few seconds ago was hard and enthusiastic inside of her, is now flaccid. John is extra sensitive after the orgasm and can't continue. They have reached the fourth phase of intercourse, *resolution*, and the blood is returning to their bodies. They lie on their backs next to each other, out of breath and sweaty.

The sperm that have landed in the innermost part of Jane's vagina have already begun the tough competition in which over 99.9 percent of them will be losers. The winner has a long and demanding journey ahead. Chemical signals in the fluid complete the development of sperm cells, a maturation process where a reshuffling of the membrane on the head and tail make it easier for the sperm to quickly swim toward their target. But where are they swimming?

In the book *Sperm Wars*, evolutionary biologist Robin Baker describes an attempt to film the inside of volunteer subjects during intercourse. The man attached a fiberoptic camera to the underside of his penis right before having sex with his female partner. The camera captured sexual intercourse from the perspective of the penis and surprised the observers in several ways. The vagina, for instance, is not an open canal or an open cylinder full of air. Instead, the moist walls rest tightly against each other, like an empty fire hose. When the man with the penis camera entered the woman, the walls glided apart, and the observers had a view of the end of the vagina as a dead end. The man removed his penis and the vagina again closed. As the couple continued intercourse, the end of the vagina became more and more like a small chamber.

From the penis camera, the researchers could see the cervix as an opening in the roof of the vagina. The man with the penis camera ejaculated and semen collected in a pool at the innermost part of the vagina.

Sperm with any hope of reaching an egg have to hurry to come into contact with the opening of the cervix. Baker uses an elephant trunk as a metaphor for how the cervix makes contact with the sperm. "Imagine that the cervix really is an elephant trunk, dipping into a large pool of semen," he writes. Only around half of the sperm that came shooting out of the man will manage to come into contact with the cervix and swim upward. The other half must go out the way they came, through the vagina.

One of the reasons why Baker pays so much attention to the cervix is because of the mucous that the cervical cells produce. He calls this "wonderful stuff." The wonderful mucous contains fibers in a network of pores that carry out a vital job, both for the woman and the future child.

The mucous of the cervix controls what gets in the uterus, and what gets out of it. Bacteria and parasites cannot enter, and menstrual blood is allowed to exit once a month. Sperm that are bad swimmers get stuck in the mucous, while quick and healthy sperm are allowed to swim freely, even though it is no easy journey. In the book *Det første mysteriet* (*The First Mystery*), Katharina Vestre describes it elegantly: "The landscape surrounding the sperm is neither clear nor hospitable. It is reminiscent of an overgrown forest, full of chaotic thickets and dead ends."

Some research indicates that cervical mucous can help select sperm pass the first hurdle. Bad sperm, on the other hand, are met with a closed tunnel. After a little while, the mucous is filled with immune cells that eat, kill, and clean up remains of the losers. Nine months later, twenty inches of baby will pass through the same well-lubricated canal.

Right or Left?

The sperm that manage to swim through the pores of the wonderful cervical mucous must then cross the uterus. The cells swim around .2 inches per minute, so in theory, they need only ten minutes to swim from bottom to top of the fist-sized organ. It is difficult to measure how long this actually takes. From experiments on women who have been artificially inseminated right before having a hysterectomy (due to disease), we know that swimming sperm are present in the fallopian tubes at the top of the uterus in just thirty to sixty minutes.

The sperm receive good help from muscle contractions in the uterus to swim upward. The muscles are the same that push out menstrual blood when the muscles curve downward toward the vagina. Around the time of ovulation, the muscles contract the opposite way, helping the sperm upward.

At the top of the uterus are two nearly five-inch-long fallopian tubes, shaped like an upside-down handlebar mustache, where the end of the tube picks up newly matured eggs from each their own ovary. The ovaries work in shifts, and each month an egg matures from one ovary, while the other has vacation. The new egg has only twenty-four hours to be fertilized, so the sperm need to be at the right place at the right time. But where is the egg? Right or left?

The sperm must make a choice, because neither ovary can guarantee the prize of an egg. Only in one fallopian tube will the lucky ones find an egg. Even if they choose the correct ovary, it is not certain that the egg will be ready. And if the egg is already on its way down, the sperm don't have much time.

To ensure that the sperm are in the right fallopian tube at the right time, the egg sends a message to the sperm through fluid that flows inside the fallopian tube. This is known from experiments with advanced x-ray imaging from the inside of the uterus of women who have trouble getting pregnant. First, researchers used ultrasound to find the ovary that was closest to releasing a mature egg. Then they added harmless radioactive clumps of the protein albumin, about the size of sperm. Over the course of the next four hours, researchers took pictures at steady intervals to see where the protein clumps had moved. One of the first things they saw was that the protein clumps had moved into the fallopian tube on the same side as the mature egg. When they added the hormone oxytocin to the blood of the test subject, the number of protein clumps increased in the fallopian tube that would soon contain an egg ready to be fertilized. The researchers concluded that the ovary containing the egg of the month controls a signal in the fallopian tubes that helps sperm choose the right path to their target.

And even if they end up in the right fallopian tube, they have not reached their target. Fortunately, sperm can survive inside the female body for a long time while they wait for an egg. Some women have reported becoming pregnant by sperm that their bodies had stored for four or five days. Where the sperm hide is still a mystery.

The fallopian tubes are a sanctuary for sperm. The vagina, cervix, and uterus attract immune cells that clean up and remove as many sperm as they can, but once sperm have made it to the fallopian tubes, they are safe. They take a break and then spontaneously start swimming before finding a new place to rest.

But where exactly do resting sperm in the fallopian tubes wait? Their break room is still a well-hidden anatomical secret in humans. In hamsters, pigs, and sheep, there are pockets in the fallopian tubes specialized for the storage and safekeeping of sperm. In some bats, sperm cells are stored for several months before they fertilize an egg. Wild boars, camels, and flies have similar storage stations, and animals are frequent guests in the labs of insemination researchers. In people, we know surprisingly little about how sperm move the last inch to the egg.

One consequence of storing sperm in the fallopian tubes is that only a few sperm reach the egg. This reduces the chance for the woman of two sperm fertilizing and destroying the egg.

Fluid in the fallopian tubes is not at rest either, providing nourishment and information to the sperm. Surrounding the fallopian tubes are two sets of muscles, sort of like in the intestines, and tiny hairs on the inside that push the fluid down toward the uterus. It's like an underwater jungle, where waves allow the egg to surf downward as the sperm must swim upstream.

If everything goes smoothly, an egg and some sperm will meet in the fallopian tube, not far from the ovary. Only about one hundred of the three hundred million sperm make it all the way here.

Bingo!

Jane is at work and is preparing for an important meeting. Inside her body, the two halves that will become a new child meet.

Around the egg is a protective and thick liquid membrane that the sperm must work their way through. The first sperm cell that reaches the egg lays its head against the egg cell membrane, and the two melt into each other. Contact between the two membranes also gives off a signal outside of the egg, preventing new sperm from following. The road is closed, and for the ninety-nine on the outside, it's game over. Winner takes all.

Twenty-three chromosomes from the sperm and twenty-three chromosomes from the egg comprise the genes of the future baby. The fertilized egg moves slowly down the fallopian tube, back to where the sperm came from, and then cell division begins. When the fertilized egg arrives in the uterus approximately one week later, it resembles a raspberry with one hundred cells. The clump clings to the wall of the uterus, sending out messages that something is happening.

When the fetus is twenty-two days old, the heart beats for the first time.

Hotwire

At sixteen weeks, the fetus begins to swallow amniotic fluid, which helps the intestines to develop normally. Later, the fetus pees in its own bathwater. The amniotic fluid also enables the growing child to move freely, protected from bumps or blows from the outer world. But there is no air in the water, so the fetus cannot breathe as normal.

The blood of the fetus does not take the usual route through the lungs to pick up oxygen, but rather takes a bypass through the umbilical cord. The umbilical cord anchors the fetus to the placenta, where blood vessels between mother and fetus are so concentrated that oxygen and carbon dioxide can flow freely from one to the other. The mother functions as an external lung for the fetus, and in addition to supplying oxygen, she contributes sugar and chemical signals through the placenta.

To achieve this bypass of getting oxygen through the blood vessels of the umbilical cord, the fetus has some elegant shortcuts in and around the heart. Oxygen-rich blood from Mom comes through the umbilical cord and into the right side of the heart. In an adult, each half of the heart is divided into an atrium and a ventricle. A fetus has the same division, but with some very important "hotwires."

The first shortcut is in the right atrium, where most of the blood from the umbilical cord cheats and passes straight over to the ventricle on the left side of the heart. Blood from the umbilical cord avoids the path around the lungs, using a hole between the right and left side of the heart instead. The oxygen-rich blood is now in the right place, ready to be pumped throughout the whole body by the left side of the heart.

In the fetus, the right atrium performs two tasks simultaneously. Used blood from the head and upper body also come into the right side of the heart, but unlike the blood from the umbilical cord, the low-oxygen blood is sent onward to the ventricle in the right half of the heart. The right atrium allows oxygen-rich blood to take a shortcut over to the left side, while low-oxygen blood is pushed onward to the right side.

It doesn't take long for the low-oxygen blood to take a new shortcut. Instead of taking the route out to the lungs, the blood is redirected to the arteries that exit the left side of the heart. From here, the blood goes through the blood vessels of the umbilical cord and out to the placenta to get new oxygen. In the book *Vital Circuits*, Steve Vogel refers to this hotwire by the fetus as "a fine bit of fluid-mechanical sleight of hand."

The shortcuts close and seal off when the baby is born, a process that occurs on its own through large changes in pressure in the umbilical cord and lungs. But fetal development does not always go as expected.

Baby in a Bag

Water in the lungs, which can be a serious symptom of acute heart disease in adults, is completely necessary for the child in its mother's belly. Fluid in the lungs help lung cells to develop normally. This development can be halted, however, should air enter the lungs.

Breathing too early can cause significant and long-term challenges for premature babies. A normal pregnancy lasts forty weeks, but sometimes the baby is born when it is only halfway developed. At just twenty weeks, it is much too early for a baby to breathe in air. Ideally, premature babies should bathe in amniotic fluid for a few more months. The technology that can save these babies can be described as a baby in a bag, to put it in tabloid fashion.

In the research lab of Emily Partridge at Children's Hospital of Philadelphia, there is a transparent container with valves and wires. On the inside is a living, unborn lamb. In the videos that followed the research article that came out in the spring of 2017, you can see a lamb dreaming and wriggling on the inside of a see-through bag.

Partridge and her colleagues use unborn lambs in their research. Lambs are normally born after 145 days, while fetal development for humans takes around 280 days. For the experiment, unborn lambs are removed after around one hundred days and then allowed to continue to develop in the bag at the laboratory. To prevent the artificial amniotic fluid from being destroyed by bacteria, the bag is a closed-off and sterile chamber. The lamb's heart pumps the umbilical cord blood through an air filter and back to the body, just as inside the mother.

The technology is about ten years away from being able to help extremely premature babies, born between week 22 and 26. Scientists emphasize that they are researching only to improve the situation for children born after week 22. But what will happen when the bag makes it possible to save a fetus that is fifteen weeks old, a fetus that can still be aborted within the limits of medical grounds?

The breakthrough with the lamb in the bag also gives a glimpse of future technology where a fertilized egg gets nine months to develop into a baby inside an artificial uterus in the lab.

For the time being, women must bear the burden of fetal development. For Jane, who is now in her seventh month of pregnancy, the fetus weighs a little more than three pounds. During the months that she is pregnant, she does not menstruate. The hormonal cycles that normally govern her body in a twenty-eight-day-long cycle are replaced by signals from the uterus. She has an excess of fluid in swollen feet, a baby that constantly puts pressure on her bladder, and an insatiable craving for soda and chocolate. Thank goodness there are only two months left.

One day, seven weeks later, she notices something wet between her legs. Her water has broken.

FOOD—LIQUID DIET

"We know almost twice as much about erectile dysfunction as we do about breast milk," American researcher Katie Hinde has said. A lack of knowledge about food and bodily fluid can have negative consequences, resulting in a feast that leaves a bitter taste: on the menu are raw milk, blood pudding, brain mass, and placenta.

The Mystery of Life

The baby in Jane Doe's belly has turned its head downward and is letting her know that it is ready to meet the world. The flood of amniotic fluid signalizes the start of a whirlwind of sensations, muscle contractions, pain, and joy. Jane does the work while John supports her as best he can. Blood, sweat, and tears, and then a baby. Little Marius greets the world with a cry, ready to discover and explore. But first, a breast with warm milk.

A baby sucking on a nipple activates pressure-sensitive nerve cells in the mother, electrical signals that go into the spinal cord and into her brain. Also entering her brain is the sound of a crying baby, which results in two things: The pituitary gland excretes the hormone prolactin, which stimulates around twenty glands in each breast to produce milk. The pituitary gland also excretes the hormone oxytocin, which activates the muscle cells surrounding the milk ducts. When the muscles contract, pressure in the glands increases, and milk squirts out of the breast into the baby's mouth.

In the period just after birth, breast milk contains around 4 percent fat and a lot of protein. In humans, the first milk is called colostrum, while the same milk in cows is called raw milk. In addition to nutrition that the child needs to grow and develop, the first milk also contains parts of the mother's immune system, antibodies that are excreted into the milk and absorbed in the baby's intestines.

Breast milk also contains living cells from the immune system, a liquid gift from mother to child. For the first few weeks, the baby is protected from infection with help from its mother's immune system, but over time, the mother's milk helps the baby develop its own mucous membranes and immune strategies. After about three weeks, the woman produces so-called mature breast milk.

Breast milk, breastfeeding, and infant formula are a billion-dollar industry and subjects of intense debate. There are many questions: How do you get the baby to suck best? Can you breast-feed at a restaurant? How many months should you breastfeed, and when is it best to stop? Will there be consequences for the baby if you don't breastfeed? How long can a woman survive by drinking her own breast milk?

Let's take the last one first: In this hypothetical scenario, the protagonist, a woman who has just given birth, has set off on a journey through a desert. The answer comes from zoologist Petter Bøckmann from the Natural History Museum in Oslo. He explains that the smartest thing to do is not to produce breast milk at all. The energy you get from drinking the milk is much less than the energy you use to make it. So, that means you will run out of energy faster if you get your body to put extra energy into producing breast milk. If, however, the milk still pushes its way out, you should absolutely drink it to ingest valuable fluid out in the desert.

Liquid Future

The World Health Organization has no guidelines for women who want to drink their own breast milk, but advises new mothers to breastfeed their babies within an hour after giving birth. For the first six months, the baby should consume only breast milk, and thereafter a combination of breast milk and other food for up to two years or longer. The Norwegian Directorate of Health gives the same advice, but specifies that accommodations may be necessary because "infants and families are not all alike."

Women who produce too much milk can earn money by donating their milk to milk banks, which send the liquid to new mothers in need. Others choose to use infant formula instead of breastfeeding, a choice that reflects what they think is best for themselves and their babies. Even though it is a personal choice, women are met with innumerable versions of "breast is best," the campaign to get everyone to breastfeed.

In Western tradition, there is heavy emphasis on baby formula being a poor and unwise alternative. The so-called breastfeeding police believe that since the nutrition that a baby receives in the first six months will have consequences for the rest of its life, women should think of what is best for the baby, not just themselves. But what do we actually know about the long-term effects of not breastfeeding?

According to some researchers, breast milk can reduce a child's risk of asthma, leukemia, obesity, and diabetes. The findings appear to be statistically robust for large groups of children, but it is difficult to translate statistics into direct and measurable consequences for individuals.

Breast milk can also make kids smarter and richer, a group of Brazilian researchers concluded in 2015. Scientists collected data from around 3500 Brazilian adults and compared the numbers with information obtained from their mothers in the 1980s. The mothers were asked about education, family finances, and how long they breastfed. Thirty years later, the children were asked about education, income, and IQ. Similar research has also been carried out in Denmark, Great Britain, and New Zealand.

The difference between children who were breastfed for under a month and children who were breastfed for over six months was almost four IQ points and eight hundred Norwegian Kroner (approximately ninety dollars) more in monthly income thirty years later. The results were published in *The Lancet*, a prestigious journal for doctors and public health researchers. "Breast milk can make the world healthier, smarter, and more equal," scientists wrote in 2016, and emphasized that breastfeeding could save 823,000 children and 20,000 mothers a year. The effects would have the largest impact on poor countries, where only a few newborn babies are breastfed for more than six months. The economic gains for public health efforts are also significant.

Critics point out that it is difficult to separate the effect of six months of breastfeeding from the effect of growing up in a rich, well-educated family. Rich and well-educated women breastfeed most often, making it extra challenging to separate the effect of breast milk from the effect of the family name. Also, are just under four IQ points crucial to an individual's quality of life?

Twin studies, the main classic in research on lifetime cause and effect, are the best way to study the effects of breastfeeding. It is understandably seldom that pairs of twins grow up in the same family, but only one of them is breastfed. On the other hand, there are plenty of siblings where one gets formula while the other is breastfed, although the comparisons are not as easy to interpret. Families with two children where only one was breastfed are of special interest to sociologist Cynthia Colen of Ohio State University.

Together with colleague David Ramsey, Colen went through twenty-five years of data of eight thousand children from four thousand families. The researchers examined the effect of formula on body mass index, excess weight, asthma, hyperactivity, relationship to parents, behavior, reading skills, vocabulary, math skills, memory, and academic skills. In the first round, breast milk did best: on average, children who had been breastfed scored better on ten of the eleven tests.

The statistical advantage disappeared when researchers compared two siblings from the same family where one had been breastfed and the other had been given formula. Colen and Ramsey believe, therefore, that the positive effects of breastfeeding come from the economic and cultural differences of the mothers, not from the breast milk itself. The debate about the impact of breast milk is not over.

Though the potential effect of breast milk may be increased intelligence, breast milk is not on the menu for adult men and women. The thought of it stirs up emotions of disgust and fear of dangerous pathogens residing in drops of milk, as author Gro Dahle discovered when she served waffles with breast milk to unwitting *Aftenposten* journalists in 1996. In a conversation with Hallgeir Opedal in 2016, Dahle said that there was probably only a drop or two in the batter, much too little to make as big a deal about it as the press did then.

The ice cream "Baby Gaga" is another culinary curiosity containing breast milk. Created by Matt O'Connor in 2011, it contained 75 percent breast milk and 25 percent cream. The milk came from fifteen donors, and the exclusive, short-lived ice cream from London was flavored with Madagascan vanilla and lemon spice.

Milk from farm animals is mostly without lemon spice, and many who grew up on a farm tell about the quality of milk that came directly from the milk cans. "The best was getting raw milk from the bull straight into the mouth—that was a hell of a lot of fun," Norwegian politician Abid Raja said to *Nationen* in a summer interview in 2017. It is the cow, not the bull, that produces milk, a biological error on the part of the politician that most of the humor columnists picked up on.

Milk Party in the Stomach

A lot of people love milk. Milk for breakfast, lunch, and as an evening snack. Warm milk with honey, cold milk after a workout, skim milk on weekdays, whole milk on Sundays.

Milk has been an important part of my culture. My paternal grandmother came from Southern Norway, was educated as a midwife and teacher, and was a big believer in the effect of warm milk on young nerves. "Drink this," she said, giving me a bottle of lukewarm whole milk before I was to compete in a music competition. Strength in the form of a drink was supposed to calm my nerves on a two-hour bus ride on the winding roads of Western Norway.

For many, milk is not synonymous with idyllic cows on Norwegian farms, but rather with stomach pain: gas in the bowel, stomach cramps, and then later, loud, wet exhalations from the rectum. They cannot tolerate milk: they are lactose intolerant.

The differentiating vowel in the words *lactase* and *lactose* are an indication to chemists that the former is an enzyme, while the latter is a sugar. In the mucous membranes of the small intestine, the enzyme lactase carries out its task of cutting the sugar lactose into two parts.

The milk sugar is cut into glucose and galactose, which the intestinal cells absorb as energy. Around two-thirds of the adult population, including most Asians, Africans, and Southern Europeans, do not have the enzyme in their intestines, allowing the lactose to pass through undisturbed.

However, in the colon, millions of bacteria are waiting, ready to take advantage of the lactose. During fermentation, gas remains. It has to get out, and the result is pain, air bubbles, and diarrhea that empty the bowel of water.

From an evolutionary perspective, adult humans were not designed to drink milk from Mom or from cows. Two hundred thousand years ago, around the start of the history of *Homo sapiens*, not all mothers were able to supply all their children with several liters of milk every single day. This resulted in children sustaining themselves on things other than milk after they were weaned.

Most children in the world stop producing lactase when they are between three and five years old. This is the norm. The exception are adults who drink two glasses of milk with breakfast every day without getting a stomach ache.

Where, then, does this exception come from? Lactose tolerance, which makes it possible for Europeans and Americans to each drink around twenty-six gallons of milk per year, is an example of newer evolution in us humans. At one time or another, a mutation arose that enabled adult people to tolerate cow's milk, and he or she had children with the same ability. The family did not get diarrhea from the milk and had a greater chance of survival than those who could not drink milk. After many generations, the mutation became the new norm. Milk for everyone. *Cheers!*

But the story is not so simple. For example, the point in time when the mutation occurred is debated, and suggestions vary between two thousand and twenty thousand years ago. Some believe that it occurred in Hungary, while others point to DNA analyses from Turkey. The truth is no simpler when it becomes apparent that there are several mutations that produce the same result: adults who can drink milk. Maybe this ability arose several times and in several places around the same time? Some think it occurred when people started keeping livestock, especially cows, while others point to the taming of camels as a central factor.

Geneticist Mark Thomas from University College London cites diarrhea and the cold as two important factors in evolution of the ability to digest milk as an adult. He explains that the first Northern Europeans were not sufficiently prepared for the short growing seasons and cold winters. When the grain harvests failed to produce enough food, many were malnourished. Cow's milk contained important nourishment, but also the risk of stomach cramping and diarrhea due to the lactose in the milk. In warmer parts of the world, lactose disappears during the fermentation process of turning cow's milk into yogurt, but nature's refrigerator in Northern Europe kept the milk fresh.

People with random mutations that enabled them to maintain a small intestine full of the enzyme lactase had a better chance of surviving cold winters with milk for dinner, without dying of diarrhea. They passed the mutation on to their children, and thousands of years later, Europe became the self-appointed king of the milk carton.

The dairy industry is under pressure, as it is not just cow's milk filling prized cartons. Some cartons contain milk with the prefixes coconut, rice, oat, soya, or almond. The word "milk" no longer means just the liquid from the breast of a mammal. As a consequence, *Merriam-Webster* now also includes the definition of milk as "a food product produced from seeds or fruit that resembles and is used similarly to cow's milk."

Recently, the dairy industry received help from the European Union tribunal. If you want to sell almond mash with water, you are not allowed to use the words milk, cream, butter, cheese, or yogurt in marketing. The same applies to products made of tofu or soya. Soya milk and tofu cheese must therefore find a new name. Advertising companies are naturally rubbing their hands in glee, writing *mylk* or *m*ylk* on the package of almond water. Game on.

Juices Flowing in the Dentist's Chair

For babies who are dependent on breast milk, it is important to get as much as possible out of it. The process, therefore, starts right inside the lips. Baby spit contains a high level of lipase, which breaks down the fat in breast milk. But spit does a lot more than transform solids to liquids.

Dentist Gro Sæther says that people who have too little saliva quickly notice the consequences. Without saliva, it becomes difficult to chew, difficult to swallow, and difficult to speak. Much of the taste of food disappears, and salty, sweet, and sour become sharp and sore on the tongue, she explains.

Sæther also points out that saliva and mucous membranes carry out an important job as a first line of defense: "Dangerous things are caught in saliva, and then swallowed." To protect against intruders, more than a few drops of spit are needed. Spit is needed in the entire mouth, all the time. For saliva to not run over, you swallow a lot more often than you think. Each day, the salivary glands produce up to one and a half liters of spit.

This plentiful liter is primarily just water with some other stuff. Saliva comes in two types, Sæther explains: the sticky resting saliva and the stimulated watery saliva. Chewing, taste, and smell can activate the salivary glands, and liquid streams out to the tongue. The mouth has about a neutral pH, but after each meal, the pH drops. Acidic food is harmful to the teeth, and saliva is there to keep balance in the fluid of the mouth.

Age, health, and life situation also influence the contents and amount of saliva. An Australian research team compared how baby saliva and adult saliva react to breast milk. They found that baby saliva was ten times better than adult saliva at reacting with breast milk and producing hydrogen peroxide, a disinfectant. A baby's spit thereby prevents dangerous bacteria like salmonella from growing in the mouth of the newborn.

An adult's spit is home to numerous bacteria colonies that cannot survive in other places in the body. Antibiotic-like chemicals in our saliva have not been found, so the potential effect of licking a wound, as dogs and cats do, has no scientific grounding at present.

What saliva does contain is traces of ancient detective stories, a field of study called genetic archaeology. In the summer of 2017, DNA analyses and computer simulations showed that at some point hundreds of thousands of years ago, the ancestors of modern Africans probably mixed with an unknown, humanlike species. The hypothesis stems from detailed analyses of spit.

As a starting point, researchers looked at one of the proteins in saliva, mucin-7 (pronounced like "s"). Mucin-7 is a glycoprotein, a protein with carbohydrates attached, that helps make saliva mucousy. The hypothesis about the unknown and extinct humans is based on small variations in mucin-7 from spit. In some Africans today, mucin-7 contains tiny variations that they probably got from someone else. Who they were, what happened to them, or what potential attributes they passed on to modern humans, it is too early to say. As usual, we must wait for new pieces of the puzzle.

As you wait, perhaps you will receive a postcard in the mail that, unfortunately, does not have inviting photos of beaches or beach umbrellas, but rather makes you start to sweat nervously. "Guess who's going to the dentist," it says in white letters across an ominous red background.

"Many are a bit nervous when going to the dentist," Sæther admits sympathetically. "Then they get dry mouth. When they eventually relax, their saliva returns." Perhaps she uses the amount of saliva to determine how the patient is doing, a sort of measurement of well-being in the dentist's chair?

Liters of Spit

At the restaurant, in any case, it is a good sign if your mouth is watering when the waiter comes. "For those of us who work with food, saliva and *juiciness* are crucial," food writer Henrik J. Henriksen explains. He is an experienced chef and is into what he himself describes as "food porn": inspirational videos and adjective-heavy texts about good food.

It perhaps comes as no surprise that he experiences an acute increase in saliva production from the classic start of almost every French dish: sautéing a bit of butter in the frying pan with some shallots. Saliva floods out of the glands right under the eye on each side of the upper jaw and under the tongue, right behind the front teeth. "All you have to do is turn it on."

Dry mucous membranes, on the other hand, result in a poor taste experience, like the taste of dry cookies and parmesan (preferably not together). At the other end of the scale are the seed pods of okra, a tropical plant in the mallow family. Okra is around 90 percent water, but is much more slimy than cucumbers or green beans. The sensation of okra slime in the mouth is "totally disgusting," according to Henriksen.

Spit does not exactly have a high status among most people. To spit on someone else is an extremely clear declaration of distrust, hate, or opposition, and drops of spit that happen to fly over the table in the cafeteria are usually followed by a hand over the mouth and an embarrassed apology. Some also think that the sounds of spit in action, such as smacking, gurgling, or slurping, are the worst noises in the world.

Food and saliva are closely connected. When the jaw is at work, the salivary glands' flood gates open, and fluid streams over the tongue to make the food malleable and easy to swallow. Saliva also contains large amounts of the enzyme amylase, which cuts the food up into smaller pieces. Amylase cuts up starch—long chains of carbohydrates in potatoes, rice, and grains—into smaller sugar molecules. If you chew for a long time, you will eventually notice a taste of sugar, even if the ingredients did not contain pure sugar to begin with.

Sugar is not just good as candy, but is also a sought-after source of energy for bacteria and yeast. When the yeast gorges on sugar, it produces alcohol as a by-product. And just like that, it has made beer.

It is a principle used all over the world, and in Peruvian rain forest drinking tradition, the alcoholic brewed beverages masato and chicha have been made with spit since ancient times. In Japan, there is a similar tradition of chewing rice to make sake. The yeast must have access to the sugar molecules in rice, yucca, or purple corn, and the starch must first be broken down into simple sugar molecules.

To kick-start the degradation process, farmers—most often women—chew on grains of rice, yucca roots, or uncooked, small, purple kernels of corn. They spit out a mash that is full of amylase. Water and spices are added to the mix before the spit and yeast change the contents into locally sourced alcohol. A *National Geographic* reporter traveling in Peru described the finished product as "a sherry-like, deep red liquid that smells like cloves and tastes like cedar."

Saliva makes it possible to preserve and store food over a long period of time. Other techniques, such as drying, salting, or preserving in vinegar or syrup, produce the same result. Bacteria growth, light, and heat can spoil fresh ingredients, posing a serious risk to health, especially liquids such as milk and blood. Also, most will be wondering where the blood comes from. Making food from blood is not for the fainthearted.

Straight from the Artery

Today I cut my finger while slicing a cucumber. It hurt, but instead of putting it under running water to wash away the drops of blood, I immediately stuck my fingertip into my mouth. Grossest is often best.

Blood tastes mostly of iron and little else. Not surprising, given that iron is important in binding oxygen to hemoglobin in each red blood cell. Nevertheless, it is an unexpected taste experience for many. Blood tastes neither slimy, thick, nor gross, but rather more natural than one would think. In addition to the taste of blood, it is almost impossible to not think about how you are blood. Your own blood, recycled.

It is not very often that I bleed. When I was served blood sausage for dinner as a child, Mom insisted that we could not grow up without having tried the everyday food—now a curiosity—of her childhood. We sprinkled sugar over the red panfried circles, once serving them with potatoes.

Grandpa's sister, Kari, born in 1924, described slaughtering animals on the family farm in the 1930s as follows:

> The pig was lying on a bench and the butcher hit it in the head, stabbed it, and read a prayer. He cut over the main artery so the blood ran out. I stirred as hard as I could. If the blood coagulated, it couldn't be used as food, so we added

salt and vinegar to prevent coagulation. Afterward, the blood had to be stored in a cold place overnight. The next day, Mom made black-colored blood pancakes. Other times she made blood pudding steamed in water and then panfried in butter. Then it was a little salty and had a crust you could bite into together with syrup and honey. It tasted sweet.

Blood pudding is an important tradition in many countries, such as the Irish variant, black pudding, which is blood mixed with grain. A Norwegian variant may contain raisins, which is where the expression *rosina i pølsa* (the raisin in the sausage) comes from, and the Spanish variant, called *morcilla* and originally from Puerto Rico, contains rice and chili pepper.

Food enthusiast Henriksen tells of an experience in Cambodia many years ago, where he was served a cup of cow's blood that they poured boiling stock into: "It almost coagulated, and with a bit of lemon, salt, and pepper, it was like warm tartare in soup-form. It tasted unbelievably good."

It may help to not think about what is in such dishes, at least if it is Hannibal Lecter serving it up. In one of the final episodes of the TV series, Hannibal makes an Italian pudding called *Sanguinaccio dolce*. The dish is a silky dessert made of pig's blood and chocolate, served in an orange peel with the fruit scooped out. In Hannibal's kitchen, on the other hand, the blood is from a woman described as "a cow, but only in a negative connotation."

With a Good Amarone

In addition to being somewhere in between liquid and solid in cooking, bodily fluids also flow on the private areas of many. When personal and bodily details become public knowledge, some react with disgust, others blush, and a few do not care at all. When the fluids that are normally inside our bodies come out, it breaks with the expectation that what is private stays private.

In another episode of *Hannibal*, series creator Bryan Fuller wanted to show two naked people kneeling beside a bed. Both were dead and had had the skin on their backs folded out like wings. The TV network, NBC, protested, not because of all the awful details of skin, muscles, sinews, and blood, but because of the characters' butt cracks. "What if we fill the butt cracks with blood?" suggested Fuller, who then got the green light from NBC.

A bloody example getting increasing celebrity-driven attention in the spotlight is new mothers consuming the placenta that follows their baby. Lucky for me, research journalist Bill Schutt has already described the experience of being a cannibal for a day.

In the book *Eat Me*, Schutt defines cannibalism as the act of an individual of one species eating the parts or whole of an individual of the same species. Schutt categorizes nail-biting and breastfeeding as a gray area, but concludes that eating a placenta is definitely a cannibalistic act.

To investigate what it is that convinces women to eat parts of their own body, Schutt pays a visit to Claire Rembis, who runs her own company of processing and selling placentas. The placenta is dried, finely ground, and divided up in small capsules. The mother takes two capsules each day, supposedly to get her energy back after giving birth.

Many species eat the placenta after giving birth, but research studies on the effect of people eating a placenta are few and lack clear conclusions. Rembis, who started eating her placenta after having baby number seven, is not bothered by this: "Consuming the placenta made me feel a bit more normal," she says. She compiles feedback from her clients, and even though she assures that she is not doing research, she admits that the anecdotes help to generate interest in placentas. "People try it, and it works for them. Then they tell their friends. It's spreading like a virus," she tells Schutt.

In the kitchen they are preparing dinner. Out from the freezer comes a bag with Rembis's placenta, cut into strips that reminds

Schutt of veal liver. Her husband, William Rembis, sautés the pieces with vegetables and a good Amarone. Schutt picks up a piece with his fork and puts the body part on his tongue. "Firm but tender, easy to chew. The consistency of veal," he concludes, adding that it tastes like other organs, or dark meat.

Recipe for a Good Cannibal

Eating the parts of other people has a long history, and in both Europe and China, traditional folk medicine has suggested eating body parts such as blood, bone, skin, and intestine to cure people, without large protest or reaction. On the islands of Central America in the sixteenth century, however, the view on cannibalism had serious implications.

Columbus and the Spanish queen labeled all the natives as cannibals, giving Spanish soldiers the right to kill them or take them as slaves. In recent years, doubt has been cast on whether Columbus and his traveling companions really observed cannibalism in village after village on all the islands and on the mainland, or if they used this as a simple excuse to take control of the new continent.

Many of the natives died of disease that the Spanish brought with them. Historian David E. Stannard, author of the book *American Holocaust*, writes that between sixty and eighty million people on the islands of the Caribbean, in Mexico, and in Central America died before the end of the sixteenth century. If the numbers are right, the word "cannibal" was the beginning of the biggest genocide in history in Latin America.

On the other side of the world in the twentieth century, cannibalistic rituals were instead met with curiosity by social anthropologists in their research studies. More and more reports were coming out of the island Papua New Guinea about a mysterious disease, described as the "trembling disease" or "laughing death." Those affected slowly lost control of their bodies and could suddenly break out in uncontrolled laughter. They shook, were eventually unable to get up, to swallow, or to breathe. They died of hunger, thirst, or pneumonia. The mystery illness killed about 1 percent of the population each year.

The Fore people were counted at thirty-five thousand inhabitants on the island, divided into 170 villages. They had had no contact with the West before the 1930s, and one of the rituals that left an impression on white explorers was the death ritual. The Fore thought it was better to be eaten by their own than by earthworms.

The ritual is described in detail by Bill Schutt: The dead were placed on a bed of leaves and the women in the family cut the body into pieces. The chunks of meat were put into piles and divided among loved ones. They placed strips of meat in bamboo leaves and roasted them over the fire before eating the meat. Finally, they opened the torso of the deceased to show the organs to the widow. The head of the deceased was also a part of the ritual, and after hair and skin had been removed with fire and a knife, they opened the skull with a stone axe and removed the brain. They mixed the semi-firm tissue with ferns before roasting and eating the brain mass. The Fore ate the entire corpse of the deceased, including the genitals and feces that had been scraped out of the intestines.

In retrospect, it is easy to see the connection between the cannibalism and the mysterious trembling disease, called *kuru*. It was mostly women and children who got sick, while adult men—who did not participate in the ritual—were more rarely affected. It was more difficult to find a medical explanation and a possible treatment for kuru.

Comprehensive research would be needed, which included injecting brain mass from a kuru victim into three apes. Three years later, two of the laboratory animals, Georgette and Daisy, developed symptoms of a neurological disorder that resembled the suffering of the people of Papua New Guinea. Georgette's cerebellum showed an identical pattern to that of the people affected: large holes in the brain tissue, like in Swiss cheese.

Kuru is caused neither by a virus, bacteria, nor parasites, but rather a protein: a prion that has misfolded. The misfolded protein is infectious because it can cause normal prion protein—of which there is a lot in the brain—to change shape from healthy to diseased. The domino effect spreads, and nerve cells die because they cannot handle the amount of misfolded contents. The brain is damaged and stops functioning as it is supposed to. Kuru resembles mad cow disease and scrapie, prion diseases that are infectious and always fatal.

I'm pretty sure that you can manage to stay away from brain mass and placenta. A traditional culinary dish that you should check out, however, contains something that is neither entirely solid nor entirely fluid, according to chef Henrik J. Henriksen. What about fried codfish sperm as an alternative to tacos?

Man Milk for Dinner

The dish has "a fantastically delicious taste," according to food enthusiast Henriksen. He says that, surprisingly, the consistency of codfish sperm, which is somewhere in between solid and liquid, does not change during cooking. Even with a crunchy crust, codfish sperm is soft inside: "It's awesome. But it's definitely a liquid," he says.

"Semen is not only nutritious, but it also has a wonderful texture and amazing cooking properties," writes a very enthusiastic Paul Photenhauer in the cookbook *Natural Harvest*. He explains that the smell and taste of semen is comparable to brie, cookie dough, almonds, salt, spice, alfalfa grass, yeast, coconut, or fruit.

Photenhauer opens the feast by admitting that he and his friends have tested all the recipes, enjoying a glass of "Almost White Russian" before a three-course meal with "Man Made Oysters," "Noodles with Special Spicy Sauce," and "Tiramisu Surprise."

Furthermore, taste, volume, and consistency are the most important factors in using this personal ingredient in cooking:

"As with all other natural, organic products, the quality of semen depends on the quality of the producer." Everyone has their own taste—literally.

One reason why many do not immediately think of blood and semen as ingredients for dinner is that these ingredients are rarely found on store shelves. Most people live a fair distance from farms or the open sea, and in the hustle and bustle of everyday life, there is no time or interest for making blood sausage from pigs and fresh cheese from cows—or exploring the taste variety of man milk.

Nevertheless, an increasing number of Norwegian consumers want products that come directly from the farm. They want to know the name of the animal, participate in the slaughter, and take a third of a sheep home from a visit to an organic farm. Customers want to get closer to nature—closer to where food comes from— and revive old food traditions. They want to make blood sausage of pig's blood, milk their own goats, and learn to eat rabbit liver. Tradition in opposition to fast food and mechanically separated meat receives good help from social media and strong purchasing power of the upper-middle class.

To get closer to nature, it is also important to know how food and drink move about on the inside of the body. Everything that comes in must go out. But where is it in the meantime?

PISS—FROM MOUTH TO TOILET BOWL

Now it's time to experiment with our bodily fluids. You need a volunteer test subject—it can certainly be you—and a big glass of beet juice. I will take part also. We are going to see how long it takes for the coloring from the beet to make our urine red. Will it take one, four, or ten hours?

Red Smoothie

It tastes sweet, earthy, and metallic, and small clumps of vegetables roll over my tongue. I've just bought a beet juice smoothie at the store. I got a lot more than beets, as the smoothie also contains carrot, banana, apple, and a few drops of pear, parsnip, watermelon, pomegranate, and spinach. It's a real vegetarian meal.

A cup of the rich beet flavor is a little too much, but I guzzle it down. The goal is to fill my bladder with red coloring, so I need to maximize my ingestion of beets. My mouth produces saliva that starts the process of breaking down the food, making the smoothie taste as it does. Gulp, gulp.

My tongue presses mashed beets, carrot, and banana back into my throat, up against the soft palate. This contact puts into motion a complex interplay of muscles that I have no control over. Commands seal off the exit through the nose, lift the larynx, plug the vocal cords, and protect the windpipe. The smoothie is sent down the esophagus where smooth muscles push the contents downward at a speed of about one to two inches per second.

Two to three seconds later, the smoothie is in my stomach. Here, the beets and other vegetables are washed in gastric acid and churned together with the remnants of my snack. Since the smoothie is already liquid, my stomach does not need to work very long. Solid food, on the other hand, needs from two to six hours of kneading in the elongated sack that begins right behind the left nipple. But what actually happens in there?

A Hole in the Stomach

A discovery from the first half of the nineteenth century gives us a partial answer, making me curious and nauseous at the same time. Military doctor William Beaumont concluded in his diary that if he pressed his tongue to the inside of an empty stomach, it did not taste acidic at all. Two hundred years ago, this oddity was at the forefront of medical research; a man pressed the tip of his tongue against another man's stomach and described the taste: nothing.

The tasteless stomach belonged to Alexis St. Martin, an eighteen-year-old French Canadian fur trapper. If you are wondering how Beaumont got his tongue down into the stomach of St. Martin, the answer is not through his mouth, but through a hole in his side. An accident with a musket left St. Martin with an open hole in his stomach. Beaumont offered first aid and discovered that his fingertip fit in the hole. Through the opening, he could see the remains of food in St. Martin's stomach.

The hole in the stomach did not heal like a normal wound. Instead, the hole grew together with the muscles and skin, resulting in a permanent tunnel into St. Martin's stomach, an opening that gave Beaumont a window to the mysteries of digestion.

Historians reckon that Beaumont hatched out his plans to experiment on St. Martin and his special stomach already after their first meeting. Nine months after the accident, the teenager St. Martin moved in with the family of Beaumont, twenty years his senior. Eventually, once he no longer needed to be taken care of, St. Martin received a small compensation for taking part as test subject in Beaumont's ambitious project.

Their first experiment started on August 1, 1825, at twelve noon. Beaumont pushed a selection of food pieces—each ingredient tied with a silk thread—in through the hole in St. Martin's stomach.

Cabbage, bread, pork, and a piece each of boiled and salted beef were attached to threads that hung out of the hole. Two hours later, Beaumont pulled out the threads, and almost everything was gone, having been dissolved by St. Martin's stomach acid. "Only the raw beef was intact," he noted. The duo continued doing research and living together for long periods over the next decade.

The results from over one hundred experiments of spying on the contents of St. Martin's stomach earned Beaumont a place in medical history. It is he, not St. Martin, who is described as "the father of American physiology." In the book *Gulp*, Mary Roach points out that St. Martin was far from an equal partner, being described as "the boy" up into his late thirties, and being forced to contribute to endless experiments, about which he understood little.

Roach explains this unappealing side of Beaumont (he was arrogant) with nineteenth-century class structure and a lack of tradition for considering the ethics of medical experiments. Obtaining informed consent from participants in research projects, something we regard as a matter of course today, was not in the minds of Beaumont and his contemporaries. Beaumont and St. Clair parted ways as enemies.

Their research results were also quite far from the truth of what actually happens in the stomach. Beaumont exaggerated the role of gastric acid and ignored both the breakdown that begins in the mouth, and the enzyme that is added in the small intestine. "The acid's main duty, in fact, is to kill bacteria—a fact that never occurred to Beaumont," writes Roach.

But silk threads with pieces of food did solve one persistent challenge of studying the inside of the digestive tract of living mammals, which is not easily accessible for tests or microscopes. The stomach hole of St. Martin was perhaps part of the inspiration for researchers when they installed holes in the sides of a small group of cows at the Norwegian University of Life Sciences.

"The temperature of the feed and the gastric juice around my gloved arm kept to around thirty-nine degrees [Celsius]," Andrea Rygg Nøttveit noted when she stuck her arm in the stomach of one of the so-called fistula cows. The animals do not suffer because of the hole, and they receive close follow-up from the researchers. The aim of Egil Prestløkken's project is to study which feed is best for the cows, how bacteria in the stomach break down different kinds of feed, and how changes in feed affect the quality of milk. An insignificant hole in the stomach thus turns out to be a win-win situation for cows and researchers alike.

In an adult person, the stomach produces between one and three liters of gastric acid each and every day. This results in a pH of about 2, an environment in which only a few bacteria, yeasts, and organisms can survive. Beaumont was mistaken when he ascribed the breakdown of fat, protein, and carbohydrates in food to gastric acid. Even though gastric acid destroys the three-dimensional form of molecules in food, it is not sufficient for getting hold of energy stored in chemical bonds. Beaumont overlooked gastric acid's converting of the incomplete enzyme pepsinogen to pepsin, one of the enzymes that cuts protein into pieces. The stomach also secretes several other important enzymes and chemicals.

Beaumont also could not explain why stomach acid does not corrode the stomach itself. Now we know that mucous membranes on the inside carry out an important job of protecting the wall. In addition to mucous, the acid-producing cells have a trick: the two parts of gastric acid, hydrogen and chloride, are spit out of the cells separately so the inside of the cell is not destroyed.

The Charm of the Badminton Court

The small intestine can also be damaged by stomach acid. Holes in the intestinal wall only cause trouble, so the intestine has several strategies for preventing the beet juice from going the wrong way and flowing into the abdomen.

The job of the pancreas is first and foremost to save the intestinal wall from the corrosive mixture of acid-washed beet smoothie. Directly after the transition from stomach to small intestine is a small opening, the end of a thin tube from the pancreas, out of which flows neutralizing liquid over the soupy mix of food.

Cells in the intestinal wall also excrete mucous to protect themselves and to help the liquid food forward. From the pancreas come enzymes as well, breaking down fat, carbohydrates, and protein, spraying the mass of liquid. The contents are divided into smaller and smaller pieces. They move onward and are cut by even more enzymes into microscopic bundles of energy that the cells can use.

Fat in food activates sensors in the intestine that open the flood gates from the gallbladder. Bile makes the drops of fat small enough to be absorbed through the intestinal wall. The canal from the gallbladder ends in the same opening as the canal from the pancreas and showers green bile over the beet smoothie.

Every day, nine whole liters of fluid travel through the small intestine. Five and a half liters come from food, drink, gastric acid, and enzymes from the stomach; three and a half liters are from the pancreas, gallbladder, and intestinal wall.

Nine liters of fluid flow through the intestine. What becomes of it all?

Most of it is absorbed already in the small intestine, where a little over seven liters of fluid follow the nutrients that the intestinal cells absorb. This principle is known as osmosis, where water moves to even out large differences in concentration of sugar, bits of protein, vitamins, and salt. The intestinal cells transport broken-down parts of the beet smoothie through itself and out to the other side before nutrients are absorbed by the blood vessels. Liquid follows thereafter to even out the differences where there is highest concentration of sugar and salt.

A great deal of cells is needed to absorb all the nutrients from the intestine, but there is not enough room for all the cells to be beside each other in the ten- to sixteen-foot-long small intestine. The intestinal wall has, therefore, thousands of small protruding, fingerlike waves also known as villi. Each intestinal cell also has its own tiny hairs on the surface, like a microscopic anemone that waves to get more food. This results in a total surface area inside the intestine that is larger than just a sixteen-foot-long tube that is one inch in diameter.

In textbooks and encyclopedias, you can read about the surface area increasing to between 1935 and 3230 square feet, popularly described as the area of a tennis court. However, in the summer of 2014, Swedish researchers concluded that the area of the inside of the small intestine is only around 323 square feet. The Swedes used advanced microscopes to calculate the effect of villi on the total area of the intestines, but found it to be no larger than a one-bedroom apartment. The total area of the intestine's mucous membranes is not the size of a tennis court, but rather closer to half a badminton court.

But the badminton court still allows cells in the intestinal wall enough time to absorb most of the nutrients in what is now a beet-and-intestinal-juice smoothie.

Fat in the smoothie, a couple of inconsequential drops, are helped along by bile, so that fat, cholesterol, and specialized protein pack together in drops. On the other side of the intestinal wall, the drops are too large to get into the blood vessels. Instead, they are caught by the lymphatic arteries.

The fat colors the lymphatic fluid milky white, and the fluid follows a lymphatic artery up through the abdomen, behind the lungs, and connects to the bloodstream right behind the collarbone. Fat from food goes straight to the heart.

To be on the safe side, most other material takes a detour. Small pieces of protein and sugar, together with salt, medications, and everything that the body thinks is dangerous stuff, goes through the intestinal wall and into the blood vessels. The cells in the body must wait for the energy, as all the blood first goes right to the liver—the most important first. The liver filters what is released into the bloodstream. Paracetamol, alcohol, and toxins are broken down, and the liver destructs old, retired blood cells as well.

The liver is the body's own detox station. If you have a well-functioning liver, you do not need to drink kale purées or ginger shots to "cleanse the inside of your body" or to "remove impurities." Your body takes care of it all on its own. The liver can also register an excess of sugar, storing some for future use.

Thereafter, newly filtered blood goes to the heart before it is hurled through the lungs and picks up oxygen. The blood then hurries out to all the cells in the body, jam-packed with fresh air, sugar, fat, and molecular building blocks that the cells need.

Wet with Tears

Cells in the body need both energy and water. The water from the beet juice flows in the arteries before moving out into the space between the cells and becoming a part of tissue fluid. The fluid passes into the cells that need it. A number of cells depend on a steady stream of fluid, for example, the glands that produce sweat or spit. Some cells are also charged with producing tears and nasal mucous at specific moments.

Increasing fluid production is the eye's response to protect itself. Tear fluid contains a long list of ingredients, and it keeps the eyeball wet and protects it against intruders. The tear glands send drops to the eye via small canals, and another set of canals transport excess fluid away from the eye and down into the nose. If drainage through the nose is insufficient, it overflows, from onions, pain, joy, or grief.

In the book *Why Humans Like to Cry*, Michael Trimble blends evolutionary history with analysis of opera and theater to investigate where our emotional tears come from. He emphasizes that humans are the only animals that experience emotional tears, an expression of an inner emotional life through flooding in the corner of the eye. Several animals can produce tears when experiencing physical pain, but only we humans start to cry from watching a sad movie, listening to music, or participating in a funeral.

Several studies have revealed that tears are accompanied by a lump in the throat, and that many feel better after crying. The cause of emotional tears is our ability to have empathy with other people, Trimble argues. He highlights mirror neurons in the brain, nerve cells that help us understand emotions in the faces of others. Mirror neurons make us feel the same emotions that we observe in the faces of others, and they explain why we get sad and cry from seeing others cry. Tears are important both for the person crying and for those watching. Tears are the most beautiful bodily fluid.

Hand Sanitizer as Deodorant

Emotions can also elicit bodily fluids other than beautiful tears. On the palms of our hands and in our armpits, we can experience what researchers call emotional sweat. An evolutionary explanation for emotional sweat is that moist hands give better grip of your surroundings if you need to flee from a dangerous situation. Sex, age, menstrual cycle, and time of day can also affect how much you sweat, as well as whether you are in the sauna, or meet a Great Dane on the loose while you are taking a Sunday stroll.

Every day, without realizing it, we lose just about a liter of water through sweat and the air we breathe out. Throughout nearly the entire body, small glands produce tiny drops of fluid. Only the lips and the head of the penis are missing sweat glands. The two to five million sweat glands are not distributed evenly over the rest of the skin, but are concentrated on the forehead and head, as well as the palms of the hands, soles of the feet, and armpits. Normal sweat glands excrete saltwater almost like a water pistol.

The armpits have another type of sweat gland, which does not function like a water pistol, but rather like a soap-bubble machine.

Small parts of cells in the glands dissolve after a while as the fluid is emptied. Armpit sweat is an oil-like fluid full of protein, fat, and steroid. But neither armpit sweat nor sweat from the rest of the skin contains odor: bacteria are the reason why sweat smells. Bacteria utilize the drops of sweat and produce waste that stinks.

So, can you use hand sanitizer to reduce the odor from sweat? Yes, says James Hamblin in the book *If Our Bodies Could Talk*. His conclusion is based on his own experiments of killing all the bacteria in armpits with hand sanitizer. He lacks a control group and objective observation methods for a clinical study, and he admits that the effect is short-lived. Deodorants, which usually contain alcohol, carry out the same task. In addition, many contain perfume, which can, to a certain extent, conceal the odor chemicals from bacteria that remain.

An alternative to killing the bacteria is to plug the sweat glands. In the store, such deodorants are labeled as *antiperspirant* and usually contain aluminum bonded with chlorine, which plugs the pores. Some research indicates that antiperspirants can reduce sweat production in the gland itself. Unfortunately, the effect is short-lived, and mixed with sweat, some versions of aluminum will also discolor clothing. So, the result of using antiperspirant to prevent rings of sweat can be rings of discoloration.

Extreme sweaters can take more drastic measures by injecting the toxin Botox under the skin of their armpits to stop the sweat glands from doing their job. The treatment can also be used on people who sweat way too much on their palms or under their feet. In the same way as in the foreheads of an increasingly large segment of the population of forty-year-olds in the Western world, Botox paralyzes important nerve signals. The effect can last for up to a year before the rings of sweat return.

In addition to fats, which are food for bacteria, drops of sweat contain discoveries that may be useful for global public health. In 2001, Birgit Schittek and her colleagues at the University of

Tübingen in Germany discovered a chemical in human sweat that can kill bacteria. Schittek named the chemical *dermcidin*, and laboratory experiments show that dermcidin can kill staphylococci, E. coli, and salmonella by making small holes in the bacteria membrane. Schittek and her colleagues believe that dermcidin can reduce the number and extent of dangerous bacteria colonies that grow on the skin of healthy people.

Bacteria are largely harmless when on the surface on the skin, but when they on occasion get into one of the sebaceous glands by the root of each hair, a pimple occurs. A pimple is a small inflammation that makes the area red and sore, but nevertheless very tempting to pop.

Teenagers—or others—with too many pimples, or pimples that are too large, often go to see a dermatologist. One of the world's most famous dermatologists is Sandra Lee, also known as Dr. Pimple Popper. She has 2.7 million followers on Instagram and nearly 3.7 million followers on YouTube. Her videos have been seen over two billion times. Hundreds of hours of yellow, brown, white, gray, and pink nastiness oozing out of the skin.

Equipped with a scalpel, tweezers, and a mobile camera, the skin doctor removes cherry-sized pimples from patients who have received local anesthesia. With what looks like the eye of a needle, she forces blackheads out of their resting places. Some of the big pimples are caused by blockages in the sebaceous glands over a long period of time, allowing the contents to collect into a dam of sorts. Other lumps are the result of harmless but uncomfortable clumps of fat. Lee also sees patients with skin cancer and makes informative and educational videos for the public.

Not everyone manages to enjoy watching videos of pimple popping, but her number of followers indicates a very large fan base. They comment, cheer, ask questions, and create fan accounts while buying T-shirts, coffee mugs, hats, and baby clothing. Fans even have their own nickname: popaholics. "Pop," they say, as yellow spots dot the mirror.

The Sneeze Lady

Small explosions of bodily fluid are everyday business for researcher Lydia Bourouiba, also known as "the sneeze lady." She is a professor at Massachusetts Institute of Technology in Boston, where she points a video camera at the mouth and nose of volunteer sneezers. The "achoo" is documented at several thousand images per second, and then researchers use advanced mathematical models to study how the cloud of particles moves outward in the room. The startling results capture the attention of many.

How far do you think drops can move when you sneeze? Most answer between three and six feet, an approximate answer based on their own experience with drops crossing the dinner table or flying from the corner of the couch over to the bowl of popcorn. Anyone who has had an attack of sneezing while eating a slice of bread also knows that snot in the nose and contents in the mouth exit at rocket speed. Slimy pieces of bread and cheese mix with snot and what you have coughed up. It is no wonder that Bourouiba refers to coughing and sneezing as "violent expiratory events."

A virus that infects the mucous membrane of the nose will cause a local inflammation. The immune system reacts by increasing mucous production in the nose and throat, as well as by creating larger openings between cells so that more fluid is released. Normal nasal mucous is clear, while a yellow-green color is the result of dead cells from the immune system. The aim of sneezing is to rinse away virus, dead mucous membrane cells, and cells from the immune system. *Achoo!*

The scientific answer for how far sneeze drops can move is both fascinating and unsettling. For the experiment, the sneeze team invited volunteers to stand in front of the camera and sneeze. Slow-motion filming showed that the drops did not all fall to the ground right away; rather, a cloud of warm, moist air appeared. The sneeze cloud rose upward, taking with it a large number of tiny drops.

The researchers concluded that the drops could travel up to twenty feet after a cough and up to twenty-six feet after a sneeze. Mathematical calculations also showed that particles could remain in the air for as long as ten minutes. The drops could also easily move up into the ceiling ventilation system and float onward from room to room and floor to floor.

Drops that fly twenty-six feet through a room and hover for ten minutes after a sneeze: the study results testify to a large potential for blind passengers to move from one person to another through the drops.

Life-Threatening Drops

It was previously believed that fluid came out of the mouth as finished drops. However, the one hundred videos of volunteers sneezing in Bourouiba's lab show something completely different. Fluid is thrown out of the mouth like flakes that expand, later developing holes and breaking up into threads of spit. The threads tumble downward while dividing into small drops, resembling paint being tossed from a can.

To study how viruses move as blind passengers in a sneeze requires seriously ill people to come and sneeze inside a laboratory that is extra secure. Researchers need to control temperature, wind, and spread of infectious drops. One of their goals is to study how different viruses and bacteria move with particles across the room, helping to establish whether an outbreak spreads through the air or from surfaces. Long-term, this knowledge can be useful in the work of containing epidemics.

Theoretical models for how drops move across a room are also useful for the building of hospitals, airports, and office buildings. If the ventilation system or airflow increase the danger of infection, one measure of improving health can be to change the building's airflow. A clear request to sneeze into one's elbow will also help reduce the number of virus particles floating in the room.

Another example of dangerous drops is bacteria that splash up from the toilet bowl when one flushes the toilet. The goal of this research is to help understand how disease spreads at the hospital, as the intestinal bacterium *Clostridium difficile* does. Researchers therefore put a toilet in front of their camera, flushed it, and then studied what kind of drops were formed and how far they could move. Could they make it all the way to the toothbrush?

Researchers had previously discovered and warned about infection via drops from the toilet, but no one had created mathematical models to understand how water from the bowl could end up in places other than the toilet. In describing how far the drops move, Bourouiba says that "it is quite shocking." The easiest way to protect yourself is to put your toothbrush in the cupboard and put the seat down before you flush.

For the one-half of the population that can choose to stand while urinating, the stream of urine toward the toilet bowl is probably a similar source of transporting drops around the room. Maybe you can feel the microscopic drops against your leg when you are wearing shorts and standing over the toilet. Fortunately, urine is at least sterile.

A Bathtub of Urine

The fluid from the beet smoothie has been around the whole body where cells have absorbed the energy and water they need. The red color of beets, which is at the core of our experiment, continues on to the kidneys. If the experiment is successful, the color from the beets will turn the urine bright pink.

The two bean-shaped organs make up less than 1 percent of the body's weight, but they receive one-fourth of the blood that is pumped out with each heartbeat. The kidneys are masters of multitasking and regulate blood pressure, sodium balance, thirst, and pH in the blood. They also make hormones and excrete waste from the body. Waste from muscle use, broken-down protein, and dead blood cells must be removed from the body, otherwise they can do a lot of damage.

When urologist Anders Debes explains how the kidneys manage all this, he states a number that I do not believe at first: the kidneys produce about forty-eight *gallons* of urine each day. This is like a bathtub full of urine. I must admit that I am amazed and impressed, but it doesn't seem to match what I have observed in my own urine production. It turns out (luckily) that almost all the urine gets absorbed back into the kidneys, so we pee only about one and a half liters of urine per day.

Debes is an expert on penises, scrotums, and prostates. I tell him about the experiment I am running and ask him to estimate how long it will take before I can expect colorful urine. A day has passed since I drank the beet smoothie, but still no sign of bright red urine. I fear that the experiment has gone down the drain and not in the toilet bowl.

Debes explains that it normally takes three to five hours before color becomes visible in urine. The liquid takes about a half hour in the stomach, one hour in the small intestine, and then moves quickly through the kidneys before being stored in the bladder.

I think back to yesterday and how after six to seven hours after the beet smoothie, my urine was more orange than usual.

The urologist explains that the color of urine is not dark red, but usually light pink. The color also varies from person to person, and the pink color must compete with the normal yellow color in urine. I can relax, knowing that I am normal, but am nevertheless a little disappointed in the rather ordinary result.

It is common for urological departments to see patients who are worried about blood in their urine. One of the control questions is whether the patient has eaten a lot of beets. Many have not thought about this, so if they answer yes, they are usually told to go home and see if it passes. Debes emphasizes that two beet slices on bread with liver pâté is not nearly enough to color your urine. Perhaps I drank too little of the beet smoothie?

What You Don't Want to Know About Urine in the Swimming Pool

It is normal to urinate between six and eight times per day, plus once at night: about one and a half liters of urine each day. Urologist Debes assures that all shades from pale yellow to dark brown are within normal range. The color comes from the yellow waste material from dead, and previously red, blood cells.

With a little chemical help, urine can match almost all the colors of the rainbow: tuberculosis medications can give urine a red or orange color, while the drug propofol, which is used in anesthesia, can give it a green hue. The pH indicator methylene blue colors urine blue, to the surprise of those who have had the substance snuck into their food. Bile can make urine the color of Coca-Cola, while coffee, saffron, tuna, and asparagus can change how urine smells.

Furthermore, the taste of urine has been of important informational value for doctors for many years. Diabetics have an excess of sugar in their blood vessels, and the sugar molecules overflow to the urine. The sweet taste of urine, which doctors studied by taking a little sip, is also why diabetes was known as "honey urine disease." Drinking your own urine to stay alive is not a good strategy, however, as your body needs even more water to get rid of the fluid in round two than you take in by drinking it.

The kidneys adjust the volume of urine to coordinate the balance of fluid in the body through sensors that react to blood pressure and changes in sodium concentration in the blood. If too little sodium or blood go through the kidneys, they produce a smaller volume of urine.

The kidneys do not operate alone, as sensors in the brain check sodium concentration in the blood passing through, sending a message if the level becomes too high. Too much sodium is an indication of too little water; then comes a signal from the brain telling the kidneys to reduce the volume of urine, as well as a signal telling you that you are thirsty.

Other times, there is a signal that there is too much fluid in the body and some of it needs to be released. For many people, this occurs via small drops that they do not even notice.

For the most part, we are in contact with only our own bodily fluids. Except at swimming pools, where unknown amounts of urine from family outings and school trips fill the pool. Urine itself is sterile and harmless, but nitrogen waste in the yellow drops can react with chlorine in the pool water and irritate eyes and lungs. The chemical reactions also produce that typical swimming pool smell. Consequently, lifeguards need accurate measurements of urine in the pool to evaluate its safety for visitors.

A Canadian research team led by Xing-Fang Li collected 250 water samples from thirty-one different pools in two Canadian cities and found traces of urine in all the samples. Their instruments were set to find the artificial sweetener E950, also called acesulfame K, a harmless and indirect measurement of the amount of nitrogen remains in the water. This E-chemical does not get broken down in our bodies, but is instead filtered by the kidneys and sent right out into the swimming pool.

Li and his colleagues measured large variations in the concentration of the E-chemical between big swimming pools and hot tubs, and between hotel and spa facilities and private residences. They then measured the concentration of E950 in the urine samples from twenty volunteers and tried to calculate how big a volume of urine there was in the pool of an average swimming hall.

The answer for a pool with a length of eighty-two feet, a width of thirty-three feet, and a depth of six and a half feet was around ten and a half gallons of urine. Even though ten and a half gallons of urine make up less than 0.01 percent of the total volume of 132,086 gallons of water, the result is clear: stop peeing in the pool!

An Experiment with My Own Bodily Fluids

I decide to try the experiment one more time. For a late dinner, I eat three medium-sized beets baked in the oven with oil and salt for forty-five minutes. The morning after: orange-pink streams in the toilet. In a moment of biological pride, I fetch my mobile phone and send a Snap to everyone I think will find joy in this breakthrough.

I am right about seven of the nine selected: not everyone can tolerate being hit with a close-up photo of pink bodily fluid at the breakfast table.

But I've been bitten by the experiment bug, and there is one bodily fluid in particular that I've been yearning to know how it looks. It will not be nearly as easy as buying a beet smoothie at the store: this time I will voluntarily stick a needle in my hip.

NEEDLE—EXPEDITION BONE MARROW

Without thinking of the consequences, I wrote to professor Therese Standal at the Norwegian University of Science and Technology (NTNU) about perhaps donating my own bone marrow to research. "We can arrange for a bone marrow aspiration, if you would like—we are happy to receive any drop we can get," she answered. Two months later, I am sitting in dread on the train over Dovre, thinking about how Standal had chuckled when she said that bone marrow aspiration is most painful for people like me: young, healthy men.

Standal researches bone marrow cancer, one of several different types of cancer that occurs in blood or bone marrow. Each type of cancer needs a different type of treatment, and NTNU houses some of the country's best bone marrow cancer researchers.

In bone marrow cancer patients, the disease affects the balance in cells that maintain bone. The liquid bone marrow attacks the hard bone. There are too many cells that destroy the bones, and way too few of the cells that produce new bone. The result for patients is that they break limbs more easily and can experience a lot of pain in the joints and bones.

Standal and her colleagues are therefore trying to develop new treatments for bone marrow cancer patients. The goal is to prevent the bones of patients being destroyed by the dangerous, liquid bone marrow. There have been many positive developments in recent years, and in the last fifteen years, average life expectancy from the time of diagnosis has risen from three to four years to seven to eight years.

To make further progress, there is, among other things, a need for understanding more about how normal bone marrow functions. I have agreed to help.

Slightly Thicker than Spaghetti

I make my way to St. Olav's Hospital in Trondheim and arrive at the outpatient clinic for blood disease where doctor and researcher Tobias S. Slørdahl has agreed to aspirate my bone marrow. He explains that bone marrow is neither gray nor transparent, but red, slightly warmer in color than regular blood.

Simplified, normal bone marrow produces three types of cells: the red blood cells that transport oxygen in the arteries, the numerous and very different immune cells that patrol around the body, and the cells that produce platelets that plug holes and stop bleeding.

Bone marrow is home to both stem cells and all the stages of development that precede the finished, mature cells that later make their way out to an artery. Some of the overviews that show everything going on side by side in the bone marrow contain over twenty-five different cells. In addition to the cells that will become part of the blood and the immune system, bone marrow houses bone cells, connective tissue cells, and fat.

At St. Olav's Hospital and other hospitals, drops of bone marrow are used to make the right diagnosis for patients with blood disease. Even though diseases can give similar symptoms, such as weakness and a poor immune system, it is important to make a precise diagnosis to select the right treatment. From a bone marrow test, blood specialists can find out what is wrong, often in collaboration with experienced pathologists. Analyses can uncover what kind of cell type has accumulated in the bone marrow, which subgroup the dangerous cells belong to, and what can be done to help the patient get better.

Slørdahl shows me the equipment he has ready on the bench. The needle has a thick, green handle and is about two inches long. The diameter is one and a half millimeters—slightly thicker than spaghetti.

The doctor draws a line on the landmark on the back side of my hip with a marker before covering it with sterile material.

I feel a little pinch on the skin, then the tip of a needle hitting the periosteum. The anesthetic is given a minute or two to work before Slørdahl asks how many fingers I can feel. I guess one, but understand by his answer that the sensors in my skin have gone on break. Only when my body starts to rock gently do I know that things are getting started.

Slørdahl aspirates 20 milliliters of bone marrow from the back side of my hip bone. The fluid does not run out by itself, but rather needs to be suctioned out into a syringe. As he suctions out the bone marrow, I feel a negative pressure in my leg, a vacuum that is uncomfortable. The pain is not localized to one spot, but distributed between my lower back, the back side of my thigh, and down to my toes. It is not very painful, but enough for me to notice it.

Later, when I listen to the audio recording of the session, it registers that less than two minutes pass from the time that the spaghetti-thick needle goes through my bone to when the whole thing is over. The aspiration itself takes seventy short seconds, a long minute where I am more preoccupied with trying to find the words to describe the pain than with actually feeling how painful it is. The pain is completely different from anything I have ever experienced before. But seventy seconds of temporary pain for a good cause is something most of us can tolerate.

When we finish, I get a bandage over the wound and a pat on the shoulder. The blood doctor enthusiastically shows me two slim vials, each containing ten milliliters of warm, dark red bone marrow straight from my body. I accept an offer to accompany my bodily fluid all the way to the experiment bench of a researcher on the third floor.

First I have to wait for a bioengineer, as the researchers would also like to have some blood. It is not all that often that they have the opportunity to draw the blood and bone marrow of young, healthy donors, so once I have signed the form, they take as much as they can get. "Now we will help ourselves liberally," Slørdahl chuckles as the bioengineer fills vial after vial.

Forty-Two Million Cells for Research

Fifteen minutes later, I am met by Lill Anny Gunnes-Grøseth, manager for the Norwegian biobank for bone marrow cancer. She thanks me for my donation and hands me a white coat, which allows me to be in her lab and observe the process of preparing blood and bone marrow for future research. Several research projects need to compare the blood and bone marrow of sick people with those of healthy people. My bodily fluids will be stored as a contribution from an anonymous donor. Grøseth records sex and date of birth on the samples, but other than that, there is no information that can later trace the fluids back to me.

Only the white cells in the bone marrow are of interest to the researcher waiting on the third floor, so Grøseth transfers the bone marrow into a special tube. This tube contains a gel that makes it possible to sort the different cell types without having to wait many hours for gravity to do the job. White cells are lighter than red, so when a centrifuge spins the tube around and around, several thousand times per minute, the cells divide over and under the gel.

A half hour later, we see the result. The bottom part of the tube is wine red with blood cells, pressed together underneath a half inch of gold-white gel. Directly above is a half-inch-thick layer of cells. The fluid is not completely transparent, but gray, as if someone has poured a couple drops of whole milk into water. Here we find all the white cells from the bone marrow. The top inch of the tube is plasma and platelets, topped by a one-tenth-inch-thick layer of fat.

Grøseth transfers each layer into its own tube before she uses a machine to count how many white cells I have donated. The machine is about the size of a washing machine but sucks in only a small volume of fluid. The cells flow through very thin tubes, past advanced sensors that calculate an answer: "Forty-two million mononuclear cells," Grøseth concludes. This type of white cell is different from others in that it has only one nucleus. Cancer patients do not always have many healthy white blood cells, so it is useful for researchers to have access to healthy bone marrow from donors.

Collaboration

Researcher Anne-Marit Sponaas sits ready on the third floor and would like to have thirty million cells immediately. I follow Grøseth out into the elevator. It dings, the doors open, and we walk down a long corridor, past ten or so offices with hard-working cancer researchers. Sponaas greets us with a smile and thanks us enthusiastically for the donation.

The four researchers closely collaborate to obtain new knowledge and develop better treatments. Although they know a lot about normal bone marrow, they do not know all the details about how the white cells in bone marrow are affected by all the chemical signals that the rest of the bone marrow produces. If research is to lead to new cancer medicine, scientists need to know more about how healthy cells behave.

Sponaas gives me a new white coat and lets me observe the start of the experiment. On a sterile work bench, she divides my cells into eight small dishes. She adds different combinations of chemical signals to the cells and crosses her fingers. For the next three days, my cells live on in an incubator at the lab.

The research project is one of many examples of how bodily fluids from donors are central to this work. Drops of bone marrow, blood, spit, and tears contain the truth of who we are, even if many of the answers are still well hidden. New research gives us important knowledge and new treatments for those who need them.

Research results also push the boundaries of what bodily fluids can tell about us.

DNA—THE TRUTH ABOUT YOU

Our bodily fluids tell a fascinating story of where we come from, and perhaps what we will die of. But what if your information ends up in the wrong hands, or gives you answers you did not want?

C.

During the First World War, the British were in need of a new way to communicate without being discovered. Invisible messages scribbled in lemon juice or pomade were no longer safe, as the Germans had already discovered how to develop the hidden codes. The breakthrough came when British agents discovered that semen could be used as invisible ink.

The fluid was easy to obtain for agents, and semen did not react with the developing methods for other invisible liquids. "Every man his own stylo," the chief for MI6—a man whose name was none other than Manfield Cumming—was reported as saying. He was also the first in MI6 to sign with just his initial, C.

In the book *Prisoners, Lovers, and Spies*, Kristie Macrakis writes that the secret agents started to "delightfully experiment with the new discovery." She adds that the agent behind the discovery was transferred to another department because his colleagues teased him so much. An agent in Copenhagen did not follow the instructions to prepare a new batch for each letter, storing his semen in a bottle instead. This led to secret messages that had a very pungent odor when they arrived.

Odor is an indication of something being not quite right, so semen as invisible ink is not a reliable winner. Heat makes the fluid change color, making the writing legible. Semen and other bodily fluids—such as urine, saliva, and milk—also light up under a UV lamp. The fluids contain chemicals that absorb the ultraviolet light and release energy as visible light. It is also possible to add chemicals that react with the bodily fluids and produce pigment that reveals the invisible messages. Our fluids are not anonymous.

Answers About Me

To try to uncover some of the secrets of my own bodily fluids, I sat on a bench in the sun one Tuesday in June and thought as hard as I could about a lemon. My saliva glands did what they were told, and sticky drops streamed down my tongue and into a plastic tube. Over the next twenty minutes, I collected almost two milliliters of spit, a surprising volume considering I needed to produce it from my own mouth without the help of water or chewing gum (hence an imaginary lemon).

Finished spitting and with a rather dry mouth, I followed the rest of the instructions in the box with "Welcome to you" written on it. I flipped the lid over the spit and screwed the cork into place. A quick phone call to a transport company, and then my spit was on its way to America. My bodily fluid was on its way to the company with the innocuous and rather inviting name 23andMe. After all, who is afraid of an international, widely discussed technology company with a name that rhymes?

Spit contains cells from the mucous membranes in the mouth and from the saliva glands; in the cells is DNA. My DNA is unique and also contains the history of my relatives, far back in time. A simple DNA analysis of a bodily fluid can therefore tell where I come from.

Or rather, it is not quite so simple, as getting an answer about where my DNA comes from depends on whether there is DNA from the old days to compare it to. There is not. There exists no thousand-year-old DNA from all the regions of the earth that can be used as a reference for where my family comes from. So, what do we do?

Well, given the lack of bodily fluids from people who lived thousands of years ago, the company uses DNA from people living today. 23andMe calls them reference databases: DNA from people with known genetic histories, usually documented through family trees and family chronicles.

23andMe compares DNA from my fluid with the DNA of the reference people. If we have similar DNA, the conclusion is that my origins are the same as theirs.

Three weeks later, I get the results about me: 100 percent European and no big surprises. I get confirmation that I am a quarter British; the rest of my genes come from Scandinavia and the region that the algorithm describes as "Broadly Northwestern European," wherever in the world that is.

In addition to geographical mapping of where in Europe my genes come from, the DNA analysis of my spit also provides information about where the different contributions are located on my twenty-three pairs of chromosomes. The chromosomes show a patchwork of heredity from my mother and father, inherited from two sets of grandparents with different genetic variants. The world has never seen my combination of genetic variation. Not yours either. Imagine that.

Since I am a man with an X chromosome and a Y chromosome, I have received a Y chromosome from my father. The X chromosome is from my mother. It is her side of the family that has British background. From her two X chromosomes, there was a fifty-fifty chance that I would be given genetic roots to the British Empire. I would not go so far as to say that I am proud of my genes, but I think it is kind of fun that I have an X chromosome from Manchester. In effect, this means that my mother and I share traits that are governed by one or more genes on the X chromosome. My better half smiles and makes a comment on genetics: "Does it apply to low blood sugar as well?"

Prehistoric Expeditions

23andMe is careful about using information from my fluid to try to say anything about who or how I am. For example, I do not have access to the health report they generate for American customers, which tells of increased risk of heart problems, diabetes, cancer, and Alzheimer's disease. I am also a little uncertain as to what they will do with the data about me, so I say no to being a part of the research database together with two million other customers. A genetic fun fact that the company can offer me, however, is that Marie Antoinette and I have a common male ancestor who lived eighteen thousand years ago.

All people who are alive today stem from the first humans who lived in East Africa a little over two hundred thousand years ago. From there, humans traveled north and throughout the world in fluctuations. Genetic variations from my maternal grandmother to my mother, and then to me, or from my grandfather to my father, and then to me, allow researchers to trace the genetic line back through history. Analyses give an imprecise picture of how my ancestors participated in the exodus from Africa.

23andMe concludes that I come from a woman who crossed the Red Sea about sixty thousand years ago. She and the man she had children with were a part of a small expedition that crossed over to the Arabian Peninsula before moving onward to Europe a few thousand years later. They were not the first ones there, however, as the Neanderthals had come to Europe long before *Homo sapiens* found their way there. Neanderthals were long described and illustrated as cave dwellers without the capacity for critical reflection or cultural expression. This is probably not true, as in more recent years, discoveries of sculptures, jewelry, and rock carvings show the close family bonds and artistic abilities of our extinct relatives. The traces they have left behind show potential for complex social interplay between individuals. In addition to fossil remains from their lives, we can also find traces of DNA from Neanderthals in forty-thousand-year-old bone remnants. We also find traces of Neanderthals in our own bodily fluids, in our own DNA.

In 2010, Swedish researcher Svante Pääbo presented the results from a large collaboration of analyzing DNA from three Neanderthals. They compared the results with five people living in different places in the world and looked for similarities. If humans had sex with Neanderthals fifty thousand years ago, there would probably be genetic traces in people alive today. The conclusion: around 2 percent of the DNA of Europeans, Asians, and other non-Africans is from Neanderthals. The contribution from our extinct ancestors is spread across all the chromosomes, and different groups of people have different genetic traces of prehistoric sex acts.

But what does 2 percent mean in the big picture? Since 2010, continually more detailed analyses have provided us with clues about what the DNA from Neanderthals may have given us. Some point to a strong immune system or adapting to a northern climate, while others blame Neanderthals for skin and blood disease.

The hypotheses are based on tiny genetic variations in Neanderthals compared to humans. If the human gene has the letter A at location number 150, and the Neanderthal gene has the letter C at the same location, 23andMe calls letter C on location number 150 a Neanderthal variant. Perhaps these Neanderthal variants can tell us something about who we are?

A Proud Neanderthal

When I log in to 23andMe, I get a message that the analysis of my fluid showed 309 Neanderthal variants in my DNA. The record is 397, and the average is 267. My eyes widen when I see that my number is higher than 90 percent of their customers. Fortunately, 23andMe is quick to assure me that the heredity from Neanderthals is under 4 percent, an almost marginal fraction of my DNA.

From a simple saliva sample, they nevertheless conclude that prehistoric sex with Neanderthals resulted in visible effects on my body. I have a gene variant that gives me straight hair, a variant that reduces the amount of hair on my back, a variant that makes me taller, and one that makes me short. The winner is a gene variant that reduces the likelihood of having to sneeze after eating dark chocolate. Thanks, Neanderthals!

Of course, gene variants are not switches that turn physical traits on and off. But there are a few simple and inconsequential exceptions, one of them being earwax. Also called cerumen, earwax is a thick and sticky secretion that protects the skin of the ear canal. The fluid comes in two types: one is sticky like snot, while the other is dry like breadcrumbs. The gene that differentiates the two is called ABCC11, a 4576-letter-long gene. At location number 538 in the gene can be either a G or an A. In the book *A Brief History of Everyone Who Ever Lived*, Adam Rutherford explains how a small change can have a large effect on earwax. People with two copies of the ABCC11 gene with letter G have sticky earwax, while people with two copies of the A variant have dry earwax. G is dominant over A, so people with one of each also have sticky earwax.

Gene variants become even more exciting when we look at how earwax is "smeared around the world," according to Rutherford. In Africa, everyone has sticky earwax, while in Asia everyone has dry earwax. Why, after humans moved eastward, did earwax become more and more dry? Researchers lack good explanations and can only speculate. One possibility is that ABCC11 does more than just change the type of earwax. Japanese scientists think that the gene variant that gives Europeans sticky earwax also helps increase sense of smell and sweat production. What role the gene plays under the arms is not known, illustrating that this simple example is a lot more complex.

23andMe explains that they have used algorithms and surveys from their customers to find a correlation and association between the five Neanderthal variants and my physical traits. The computer looks for situations where different things occur simultaneously, such as a gene variant and amount of back hair. The result is then compared with the average person. The effects of one gene variant or another are tiny. Nevertheless, less back hair and silent gorging on chocolate? Yes, please.

Can we also use gene variants to explain who we are and why we do as we do? In 2009, an Italian judge had the same question, and his answer surprised many. The case involved an adult male with psychological problems who had first been sentenced to nine years in prison for killing another man. During the appeal, the defense attorney arranged a DNA test for gene variants in four different genes linked to violent behavior. The defendant tested positive. Among others, he had a gene variant in the gene MAOA, which other researchers had previously found in violent and criminal men.

The judge believed the defense attorney's explanation that the gene variant made the defendant more inclined to commit violent acts. His sentence was therefore reduced from nine to eight years, a decision that prompted large international debate. Using gene variants to explain why people do what they do is closely connected to the debate surrounding free will. Is it I who decides, or is it my genes? A potentially more serious consequence for people with these gene variants is that society expects future crime and increases—rather than reduces—prison sentences.

Ten Percent of Africans

A deep dive into genetic prehistory does not always give such simplified answers as with my saliva sample or the Italian murder trial. Long before we knew about DNA, blood determined kinship and social stratification. Color of blood was important, where royalty—those with blue blood—were above red-blooded commoners. Blood brothers were more than just friends, and pressing their bloody thumbs together made people more than just regular siblings. Amount of blood was also important, and in the American southern states, it took no more than a great-grandparent of African origin for a person to be accused of also being African.

One drop of African blood in the veins excluded a great-grandchild from white society. The blood in one's veins signaled family history, attributes, and qualities.

Now anyone can replace his or her family tree with a DNA test where a saliva test or a drop of blood is all that is needed. For some, the goal is to prove to themselves and others that they belong to a certain group. The result is not always as expected. What do you say when you are a white nationalist from the Alt-Right and you get a letter in the mail saying that you are 10 percent African?

According to Aaron Panofsky and Joan Donovan from the University of California, Los Angeles, white nationalists react in three different ways when they receive such news. Researchers have studied over three thousand posts on Stormfront, an internet forum established by a former Ku Klux Klan leader at the end of the 1990s.

In their study report, Panofsky and Donovan separate the different reactions to what nationalists understand to be "bad news." The nationalists' goal is to minimize the fallout of the DNA test on their lives and ideas. Some choose to discredit the test itself, going as far as to claim that the company that did the analysis is part of a Jewish conspiracy. Instead, they build their identity on traditional family trees undisturbed by modern technology. Another group questions the statistical methods and gives alternative explanations for how their DNA can resemble the DNA of people who live outside of Europe.

A third group of the internet forum users appear to accept the DNA results about themselves. One consequence of this is discussions where users attempt to expand or change the definitions of what it means to be white. They ascertain, among other things, that if the definition of white is based on strict interpretations of DNA tests, nationalists will run out of new members.

The vast majority of people have genetic histories that tell of mixing between different people from different places of the world. Obtaining information from a couple of drops of fluid can be important for understanding our identity as adults. For children, obtaining knowledge from bodily fluids can be a matter of life and death.

Eternal Baby Drops

In the book *Norske forskningsbragder* (*Norwegian Research Exploits*), Unni Eikeseth tells the stories of six-year-old Liv and four-year-old Dag. Their parents had noticed early on that their children were not developing normally, and that there was "a strange odor from the child, from both the urine and breath." Both children were severely developmentally handicapped and could neither speak nor move. In January 1934, after many failed visits to the doctor, the mother took Liv and Dag to professor Asbjørn Følling at Oslo University Hospital.

Følling tested the children for diabetes by adding a few drops of iron chloride to their urine. He waited for a purple or burgundy color, but the urine was olive green, a result Følling had never seen before. For the next months, Følling used his biochemical knowledge to find out what colored urine green. At last, he found himself with a flask of white crystals of the chemical phenylpyruvic acid. After doing more research and tests, Følling concluded that Liv and Dag had a serious metabolic disorder. Their bodies were unable to break down the amino acid phenylalanine normally, so the breakdown product phenylpyruvic acid was excreted in the urine.

Today we know that an enzyme in the liver has the task of breaking down phenylalanine. Over four hundred different changes in the enzyme can cause the disease phenylketonuria, also called Følling's disease or PKU. Although there is no cure, children can have a good life by following a special diet that does not contain large amounts of phenylalanine. The metabolic problem can lead to brain damage if the disease is not discovered early.

Uncovering serious disease through drops of blood is part of a baby's very first meeting with the Norwegian health system.

Through screening of newborns, a small drop of blood from the baby's heel is sent to Oslo University Hospital for testing. The Directorate of Health offers all newborns testing for hearing loss, phenylketonuria, and hypothyroidism. Testing can be expanded to check for over twenty rare, congenital, hereditary diseases that are important to treat as soon as possible. From the fall of 2017, two diseases that seriously compromise the immune system were also included in testing.

Since all newborns in Norway are screened, the blood samples can be a very useful, but also controversial, biobank of all Norwegians. The rule has been to destroy the blood samples after six years, but in the fall of 2017, the government proposed long-term storage of blood samples from newborns. Blood samples in the biobank are to be used for "healthcare, quality assurance, method development, and research." In practice, the biobank will give researchers access to analyzing DNA from all Norwegians without having to obtain consent for each specific project. The Biotechnology Council, the Norwegian Medical Association, the Children's Ombudsman, and the Norwegian Data Protection Authority have all warned against long-term storage of blood samples.

For what if someone carries out DNA analysis of all the samples in the entire biobank? Then they suddenly have detailed information about all Norwegians, information that is personal and possibly sensitive. In China, parents can pay fifteen thousand Norwegian Kroner to analyze all the genes of their newborn baby and get information about risk of disease, as well as a selection of physical traits. These registries can also be used by the police, the state, pharmaceutical companies, and insurance companies. It becomes very difficult to maintain control over who has access to the data, and it can become difficult to make sure that no one is subjected to genetic discrimination.

At the same time, storing DNA information about citizens is a very useful tool. The police have three different types of DNA registry, collections that have a part in different types of investigations. In the identity registry, the police have an overview of almost seventy thousand people who have been sentenced to prison, detention, or community service, as well as some people who are volunteers or who have received fines. The registry does not contain complete DNA analyses, but is the result of a targeted investigation of gene variants at seventeen locations in the DNA where there is large variation between people. The result is a DNA profile, a unique genetic fingerprint that differentiates people from each other.

The temporary investigation registry has DNA profiles from over ten thousand people who are suspected of committing criminal acts. The tracking registry contains an equally large number of DNA profiles where police store all unidentified traces of DNA from crime scenes.

Would it not be a good idea for detectives to compare unknown DNA profiles with blood samples from newborn screenings? And would it not also be useful for researchers to have access to an enormous biobank to find treatments for rare and serious diseases? The answers are not simple, and the conflict between the right to privacy and the greater good of society fuels tension and discord.

The compromise between the interests of research and privacy that members of the Biotechnology Council suggest is allowing storage for up to sixteen years. By then, a child is of legal age to make health-related decisions and can decide whether he or she wishes to be a part of the biobank.

In contrast to newborn babies, adults can decide for themselves what health information they want to receive from a small drop of bodily fluid.

One, Two, or Three Dots

Within just a couple of minutes, analysis of bodily fluid can give important answers about the health of a person, about what is wrong, what the consequences will be, and what can be done. Small drops give big answers.

I take the elevator up to the Health Committee's offices in downtown Oslo to take a so-called rapid test for HIV. A smiling face opens the door and shows me to the waiting area.

On the table are cookies and condoms, and on a shelf are brochures about safe sex and all the different sexually transmitted diseases. A young man, one of the volunteers who takes the tests, takes me to a room with two yellow reclining chairs and some big potted plants.

I sit down on one of the chairs while he takes out his equipment. Wearing gloves, he opens a sterile bag and pours its contents into a cup made of recycled cardboard.

An HIV test is easy enough to carry out that it is accompanied by a similar test for the bacteria that causes syphilis. The rapid tests check the immune system's response to infections in the body. In 2016, the Health Committee carried out just under two thousand of these anonymous rapid tests.

The man points to a white egg cup with a square bottom where the test result will become visible a minute later. One dot at the top to indicate that the test works, one dot at the bottom if I test positive for HIV, and one dot on the left if I test positive for syphilis. If there are three dots, I have tested positive for both HIV and syphilis.

He takes my middle finger, takes a needle, and lets me know that there will be a small pinch. I don't feel anything, but a drop soon trickles out of my fingertip. With a small pipette, he suctions up a drop of blood before sliding the pipette into a little bottle. The liquid is colored orange-red by my blood.

He pours the orange liquid into the egg glass where it quickly goes through the filter. Then he pours in a blue liquid from the miniature bottle with a number two on it, and finally, a clear liquid to wash the filter. I am prepared for sixty long seconds of embarrassing quiet and uncertain waiting, but before I can change my mind, he has reached his conclusion: one dot, everything is fine.

Separating the Healthy from the Sick

In the bodily fluids of blood, saliva, and urine, simple tests can find signals that a person is sick. A pinch on the finger can find antibodies for HIV, and a pinch on the heel of a newborn can find high levels of phenylalanine. Progress is constantly being made in the research on signals of disease, signals often described as biomarkers.

Biomarkers are biological units, like cells, protein, or DNA, that change levels in a bodily fluid when the body is sick or not working as it is supposed to. Blood sugar is a biomarker that tells how the body responds to nourishment, while the chemical PSA from the prostate gland helps to determine whether a man has cancer. If it is possible to measure blood sugar and PSA levels in bodily fluids both quickly and reliably, simple tests can help get the right treatment to those who need it.

There is stiff international competition among producers and researchers, and the tests are still a ways away from doing what they promise. Because the technology is very advanced and the bio-markers are many and complex, there is large room for uncertainty.

Astrid Aksnes is comfortable with a bit of uncertainty. She is a professor at the Institute for Electronic Systems at NTNU and currently leads an interdisciplinary research group developing what they call a "Lab-on-a-chip": a laboratory that is just one square centimeter in size. Within twenty minutes, this little gadget is supposed to give doctors lifesaving information from a drop of bodily fluid.

Aksnes shows twenty small squares of silicon shining in black and dark green. One of them has thin plastic tubes connected to it that transport bodily fluids over the surface of the silicon. On the bench, one of the researchers has prepared a chip for testing. On the left is a laser, and on the right, hidden under a cardboard box, a detector. Fiberoptic cables guide the light from the laser over the chip, where the lights will meet the fluid, and onward to the detector, where the computer is notified about biomarkers in the drop.

As long as biomarkers are correct and specific, Aksnes's little instrument can measure the levels of several biomarkers at the same time. She imagines the chip being used at doctors' offices around the world to give quick answers to important health-related questions. The doctor puts a drop of fluid from a patient onto the chip and twenty minutes later they receive the result they need.

The plan is for the prototype from NTNU to find three different biomarkers simultaneously. A sensor for C-reactive protein (CRP) can detect traces of inflammation, while a sensor for lipocalin-2 indicates that the kidneys are not functioning normally. The third sensor is for TNF, a signal that can detect both inflammation from rheumatoid arthritis and the development of certain cancers.

The three biomarkers are not only supposed to indicate whether a patient is sick; Aksnes has also chosen these three biomarkers to challenge the development of technology. If the technology works, the chip can later be expanded to include dozens of different biomarkers for different diagnoses. Aksnes is in close dialogue with international researchers working on similar projects, and she is open about the fierce competition in the development of new technology for discovering signals of disease in bodily fluids.

Periodically, we can read enthusiastic press releases about new tests on bodily fluids that can easily uncover a handful of dangerous and scary diseases. The truth is not always so easy, because it is no simple task to sort out a particular signal that can give reliable results.

The biggest challenge has to do with separating the healthy from the sick, as no two people are alike. Sex, age, and genetic variations make it extra difficult to find small nuances in individuals, though field professionals are getting closer to technology that will make it possible to make a very precise diagnosis from a single cell in the blood.

But not everyone keeps the promises they make.

The Theranos Saga

In a press photo, Elizabeth Holmes is dressed in a black turtleneck sweater. She reminds you of the world-famous founder of Apple, and in the spring of 2014, she was constantly being described as the next Steve Jobs. Positive articles and big promises filled all the major American newspapers. Her idea was no less than a medical revolution: one simple blood test would give consumers quick results.

Holmes started the company Theranos as a nineteen-year-old in 2003, the same year she dropped out of Stanford. The year after, investors contributed 6.9 million dollars, and within ten years she had collected over four hundred million dollars. The company was to produce diagnostic tools for the pharmaceutical industry and went with cheap, quick tests to determine dosage and effect of pharmaceuticals under development.

The company's value was set at nine billion dollars, and in 2015 it contributed to a change of law in Arizona that allowed private customers to test their blood without a doctor's approval. Several large mall chains signed collaboration agreements with Theranos, and in the stores, customers were offered a blood test in one corner, with a sign directing them to the shelf of prescription-free medications in another. The fairy tale stormed on, putting everything in place for customers across the entire USA to get access to simple and safe blood tests with useful health information.

But soon thereafter, the façade began to crack. *New York Times* journalist John Carreyrou was skeptical about the secrecy surrounding the technology that Theranos used to analyze the blood samples. Since tests for blood samples do not require approval from the authorities, Theranos was able to develop and use new methods without involving international experts and critics. Carreyrou started digging and eventually made contact with central informants. A few months later, he published his first article.

According to Carreyrou's sources, the technology that Theranos used was not up to par. Their machine, called Edison, was unable to find adequate amounts of signals in the blood. A central problem was that blood in small amounts did not behave the same way as large amounts, and blood cells flowed differently through the machine. Technicians had therefore diluted the blood samples and used standard equipment produced by Siemens to deliver the test result.

Holmes defended herself against criticism by calling the newspaper article an attack on innovative entrepreneurs who wanted to change the world. She was unable, however, to provide satisfactory answers convincing that everything was as it should be. Six months later, it was revealed that an inspection at the Theranos laboratory had uncovered serious errors and inadequacies. Furthermore, it was revealed that the company had used a defective test for blood coagulation, and that incomplete test results had influenced medical decisions. The inspectors threatened sanctions and the company's value fell from nine billion to eight hundred million dollars.

Finally, in the spring of 2016, it was revealed that Theranos had thrown away and annulled all the test results that the Edison machine had delivered in the period between 2014 and 2015. Additionally, the company sent out thousands of letters to doctors and consumers where it informed about and corrected test results. A few months later, the company lost its approval to analyze blood tests from customers insured through Obamacare. Holmes was also banned from running a laboratory for two years—a serious blow to an entrepreneur with high ambitions.

Several months later, Holmes announced that the company would be laying off around half of its 750 employees and stopping analysis of blood tests. For anyone other than Holmes, this defeat would have been the nail in the coffin of the dream of changing the world. But she was not going to give up.

Holmes has instead bet on medical equipment for the pharmaceutical industry. Now it's all about the platform miniLab, a box the size of a shoebox. Holmes has stated that the new goal of Theranos is to commercialize small and automated laboratories that can test a small volume of a bodily fluid. The company will develop equipment "with a focus on vulnerable patient groups, infants, and patients in need of emergency medicine." Astrid Aksnes at NTNU has a new rival.

Am I Going to Get Sick?

There are also fluid tests that offer a glimpse into the crystal ball. To a certain extent, the contents of the fluid can tell something about the future, and for some, the results have major consequences.

"We can do genetic tests on healthy people and know for certain whether they will get a disease or not," says Ellen Økland Blinkenberg, geneticist and attending physician at Haukeland University Hospital in Norway. She works with families with serious genetic diseases and tries to make complex medical research accessible so that the patient and family have enough information to understand the situation they are in.

The test she is describing is for Huntington's disease, a dominant gene defect that is inherited from one parent. Those affected usually get sick as adults, and the brain disease causes involuntary movements, loss of physical abilities, and development of dementia with altered personality. As of yet, there is no treatment that can stop progression of the disease.

If one parent has the disease, there is a 50 percent risk of each of the children also being affected. The test gives a certain answer as to whether the person will get sick or not, and the result can help people plan their own lives, prepare their family, and decide whether or not to have children.

Blinkenberg highlights an informal survey among members of an American patient association. Before the test was available, almost all were positive that they wanted to get tested. But after it became available, only around 20 percent of those with a fifty-fifty risk took the test, so the demand for genetic testing for Huntington's disease is not as large as one would think. Blinkenberg believes a conscious decision lies behind not getting tested, as "the test interferes with lives and life choices."

Biotechnology laws ensure that people who wish to take such a genetic test through the Norwegian healthcare system receive genetic counseling in advance. Private customers can, of course, purchase tests from the internet, such as from 23andMe.

Tailor-Made Red Wine

"Do you really want to give them your genes?" Blinkenberg asks me when I tell her about the saliva sample I have sent in to 23andMe. She is relieved when I say that I declined to participate in the research database of the American company.

23andMe started up in 2008, and just a couple of years later, they offered personal genetic testing for 254 different diseases and conditions. The results included comprehensive mapping of health risk and response to a long list of pharmaceuticals. The internet site gave consumers the impression that they could safely use the information to make informed decisions about their health and to prepare treatments. Inspectors from government authorities were not convinced, and since 23andMe lacked approval for testing, they were strictly forbidden from genetic testing of saliva samples for providing answers about disease risk.

Since 2013, the company has worked diligently to get documentation in order and to secure procedures to once again be able to provide consumers with DNA-based health data. Now they are back with new promises. In the spring of 2017, 23andMe obtained permission to sell their customers a health report with the results of ten specific genetic tests. The health report contains genetic tests for Parkinson's disease and Alzheimer's disease; celiac disease; two diseases connected to blood coagulation; two diseases connected to red blood cells; one disease that causes problems with the liver or lungs; one disease that causes involuntary muscle activity; and, finally, Gaucher disease, a genetic disease that can affect and cause symptoms in the whole body.

The test for the ten diseases is only available in the USA, so I will not get to know about my own future without taking another route. 23andMe is not allowed to give me health information based on my saliva sample, but I am free to sell my data to other companies that will gladly carry out all sorts of genetic tests.

There are no limits to genetic testing in the private market. I can get detailed results on future health, how I should adjust my workout to my genetic attributes, which foods I may be allergic to, what I should eat to be healthy, how my body breaks down common medications, and which narcotics I may become addicted to. Not all of them are equally serious, and some of them use the wow factor of highly technological genetic testing to market a rather standard questionnaire.

As an example, the company Vinome offers a genetic test to find out what kind of wine I like best. The analyses are in collaboration with one of 23andMe's competitors, called Helix, which carries out the DNA analyses and sells the results to the company, which then interprets them for me. Vinome offers tailor-made vintage wine. Their qualified guessing is based on analysis of ten gene variants in nine genes involved in taste and smell. They also ask for detailed information about their clients' food preferences: What kind of coffee do I drink? What do I think about cheese? What about coriander and Brussels sprouts? My self-reported taste preferences probably give a much better idea of what kind of wine I like than the nine selected genes.

Vinome combines responses on the questionnaire with the DNA analyses, concluding in a taste profile of the wines I will like—wines that I can easily obtain through an exclusive membership in their wine club. For six hundred dollars, I can receive three bottles of wine each quarter for one year. My DNA results give me access to hidden pearls from exclusive vineyards in California, they promise. This can seem like innocent fun, but for companies that replace red wine with pharmaceuticals or health foods, the consequences of taking results on good faith can be much larger. It is probably wise to be skeptical about genetic tests that give simple answers.

Geneticist Rob Arthur is also skeptical, and in 2016 he sent his saliva sample in to 23andMe. He then sold his DNA results to a handful of private companies and wrote an article in *Slate* magazine about some of the surprising health advice he received. For the most part, their conclusions cannot be trusted, and according to Arthur, companies offering "nutrition genetics" offer the least reliable information. Their conclusions about vitamin pills and supplements break with scientific consensus because scientists do not yet have sufficient or solid enough documentation. "Without a doctor's expert advice, consumers can misinterpret such advice with dangerous effects," Arthur writes.

The Postman's Son

That is to say that there are limits to the precision that can be attained in analyzing a person's bodily fluid. In only a minority of genetic tests is there a clear connection between gene variant and changed biochemistry in cells. The vast majority of attempts to draw conclusions stem from data about correlation. Things happen simultaneously, but we do not know why. It is not certain that one is the cause of the other. The results of a genetic test from a saliva sample must therefore be taken with a large grain of salt. So, is there anything to be afraid of?

Blinkenberg is worried that increased use of genetic testing will give people more health-related anxiety, and that people spend a lot of money on buying themselves worries. The two thousand Norwegian Kroner I spent on my saliva analysis is a good example.

Any possible worries I have will also affect my family. I share 50 percent of genes with my siblings, and if I learn about something surprising and dangerous in my DNA, it could force them to decide what they themselves want to do. These decisions can also have consequences for family planning and family relationships. What if I had discovered that I was the postman's son?

Many consumers are worried about the risk of future disease affecting the price of their health insurance premiums, but for the time being, insurance companies are not allowed to save or use information from genetic tests. The US has similar protection laws against discrimination based on genetic information, but it is uncertain how things will be in the future. Blinkenberg believes, for instance, that a saliva sample will be a part of the process of purchasing insurance in the future, a scenario that the American company GWG Life has already implemented for their customers of life insurance. On Twitter, experienced scientific writer for *The Atlantic*, Ed Yong, comments that a saliva sample "is like reading slightly less-reliable tea leaves."

"Curious people can do as they will," says Blinkenberg. I have no need to know what I'm going to die of.

Some, on the other hand, will do almost anything to live longer.

EVE—THE DREAM OF ETERNAL LIFE

If you are over thirty-five years old and have eight thousand dollars to spend, you can be part of an experiment that might delay your death. All you need to do is take a drive to a private clinic in Monterey, two hours south of San Francisco, and get injected with two liters of blood plasma from an eighteen-year-old. According to the man behind this controversial treatment, Jesse Karmazin, the goal is to test what young blood does to volunteer test subjects. The firm is named after the food that made the Greek gods immortal: ambrosia. A modest homepage contains just three words: "Young blood treatment." The dream of an elixir of life is far from dead.

Dirty Talk

Norwegian sitcom character Marve Fleksnes would like to do something for his country, and he has the choice of either joining the political party Venstre or giving blood. In the episode "Blodgiveren" (The Blood Donor) from 1972, we see Rolv Wesenlund in the role of a self-sacrificial fluid donor. It is his first encounter with the blood bank, so he offers his left wrist to the nurse, played by Kari Simonsen, and smiles broadly: "Go ahead!"

In the waiting room, where the walls are decorated with milk, dentist, and nutrition advertisements from the early 1970s, two other blood donors wait their turn. Fleksnes is anxious, and after a lot of hemming and hawing, he has his blood drawn. Later he calls back to check who got his blood, and whether it was someone who deserved it. The answer he receives is short and dismissive: that's not how the blood bank works.

Fleksnes is upset and distracted when he attempts to open a jar of jam with a sharp knife. Inevitably, he ends up with a big cut on his hand. The ambulance transports him back to the hospital where he is to receive a blood transfusion. He ends up having an unusual blood type, but luckily, the blood bank has exactly half a liter in stock. Fleksnes ends the episode with the following: "Gee, the blood bank is nice. You make a deposit when you have a little too much, and then withdraw it when you need it. Yes yes, nurse. Get started. Fill 'er up. Full tank, please."

One fall day in 2017, there is a lot of activity at the blood bank in Bergen. It is the week before the World Championships in cycling, and on the information screen is a notice that they "are working as quickly and responsibly as they can." On the table in the waiting room are stacks of Vestlandslefse pastries, crispbread, and muesli bars, as well as iPads and newspapers supplementing the brochures thanking you for being a blood donor. A double-door refrigerator supplies donors with Pepsi and Farris mineral water, and the coffee machine hums steadily.

Einar Kristoffersen, professor and manager of the blood bank, begins the tour by showing me the digital health form that all blood donors must fill out. A digital tablet asks for a report on your health, sexual habits, and vacation travels since you last gave blood. The questions are direct, not unlike the intimate information that Fleksnes referred to as "dirty talk."

If you have a piercing in a mucous membrane, you cannot be a blood donor. Kristoffersen explains how an innocent nose ring breaks the protective mucous membrane. At the blood bank, the hole is considered to be much more than decoration, as the titanium ring leads to a small but chronic infection. This results in a tiny chance of bacteria crossing from the surface of the skin into a blood vessel. So Kristoffersen says, "Thanks, but no thanks" to blood donors with nose rings.

Donors that vacationed in Tuscany this summer are also turned away. They receive a quarantine of four months because local authorities discovered a few cases of West Nile Virus in the region. Those who have had an autumn cold cannot donate blood either. The rules are strict, and those who are allowed to donate their blood are healthier than average. "I don't know if they know how lucky they are," Kristoffersen comments about blood donors who donate twice a year, year after year. After four visits, they receive their first Moomin mug in exchange for their donations.

Donating blood is not a contest. What was a contest, however, was the race to be the first to carry out a successful blood transfusion between two people. This competition was hard on the animals of doctors who tried to outdo one another's experiments.

Animal Blood

In 1665, nearly forty years after William Harvey discovered that blood was pumped round and round, an English doctor transfused blood between two dogs. The following year, a similar experiment was carried out in front of members of the British Royal Society academy of science, who observed how blood flowed from a mastiff into a spaniel. This prompted scientific excitement among members, even though the mastiff bled to death from the experiment.

In the summer of 1667, Frenchman Jean-Baptiste Denis transfused blood between a sheep and a fifteen-year-old boy. Denis had obtained a bachelor's degree in theology, become a qualified doctor, and completed a doctorate in mathematics. At twenty-seven years old, he was employed as a professor of mathematics and astronomy in Paris. But his hobby was medical research, and Denis's breakthrough came when he met a teenager with prolonged fever. The young boy was exhausted after more than twenty bloodletting treatments with leeches.

Denis concluded that his problem was too little blood, so he initiated an experiment to supply the boy with more of the life-giving fluid. Denis connected a tube between the sheep's carotid artery and a vein in the boy's arm, and let the high pressure from the artery push the sheep's blood into the teenager. The boy survived probably because the volume of the foreign sheep's blood was quite small, about three hundred milliliters. Denis attributed the success to the living sheep's blood, although stopping the bloodletting with leeches was probably at least as important.

Six months later, Denis tried to use blood transfusion to calm down a man who was causing a ruckus in the street. Denis had previously transfused blood to a healthy man with great success, and to a sick man with less success—he died right after the procedure. This time, the goal was to calm the agitated man, Antoine Mauroy, with blood from a serene calf. Mauroy died during the third transfusion, and his widow accused Denis of murder. The French judicial system acquitted Denis, but at the same time banned blood transfusions that were not under the direction of the French medical faculty. A few decades later, both French and English authorities banned blood transfusions completely.

A hundred and fifty years would pass before the first successful attempt of blood transfusion between two people. In one of the first experiments of British obstetrician James Blundell, he transfused one deciliter of blood from husband to wife to stop bleeding after a birth. Through his work, Blundell could see how serious it was when women lost a lot of blood while giving birth. To combat the helplessness he felt, Blundell carried out ten blood transfusions in the period between 1818 and 1829. According to historian Chris Cooper, "Over half of the patients died, though some were probably very close to death (or actually dead) prior to the transfusion."

Red, Yellow, Green, and Blue Blood

At the blood bank, a woman sits in a gray recliner, reading a magazine while she squeezes a ball. A tube goes from her left arm into a machine the size of a small washing machine; the noise is an indication of something inside spinning round and round. Two bags, neither of which are red, hang above her chair. Einar Kristoffersen explains that the woman is donating only platelets and plasma; her red blood cells are immediately returned to her.

By centrifuging the blood, the contents are divided by weight and size: red blood cells on the bottom, platelets in the middle, and plasma on the top. Some donors need all the red blood cells for themselves, but donate plasma and platelets to the community.

The two transparent plastic bags are yellow on the inside. I expected that they would be the color of white wine, like a straw-colored Riesling; instead, the bag of plasma is the color of apple juice from concentrate. The yellow color comes from the same thing that gives urine its color: color remnants from dead and broken-down blood cells.

"We know very quickly if the donor takes birth control pills because then the plasma is green," says Kristoffersen. "Some medications can also make plasma pink," he smiles.

In the back room, I put on blue plastic slippers over my shoes before meeting two bioengineers. Their job is to "produce the blood," a technical term they use about examining and preparing the blood to be used at the hospital, not the task that the donor's body has carried out. Unique bar codes keep track of the donor and all the bioengineers that have touched the bag all the way from "Hello, welcome to the blood bank" to when a stranger somewhere else at Haukeland Hospital receives a transfusion. The engineers must adhere to very strict rules of monitoring, because even though blood saves many lives, it can also cause serious injury. Blood produced by the blood bank is therefore subject to comprehensive testing to uncover viruses and bacteria.

A special type of blood can also test the safety of laboratory-produced vaccines, insulin, and intravenous fluids. The fluids can contain toxic traces of bacteria, so to test whether they are safe to use, liters of blood are needed: not regular blood from humans, but light-blue blood from the ocean.

The blue blood comes from horseshoe crabs. Even though they live under water and look like crustaceans, horseshoe crabs are not real crabs, but closer in relation to spiders. Every year, horseshoe crabs are gathered on fishing trawlers or picked up from the beach, have their blood drawn, and are released. Whereas iron and hemoglobin make the blood of humans red, copper and hemocyanin make the blood of horseshoe crabs blue. Blue blood is special not only for its color, but also for a chemical on the inside of the horseshoe crab's blood cells.

In the blood cells of the horseshoe crab is a chemical that produces a clump of jelly if it comes into contact with bacteria toxins. No clump means no danger. The light-blue test is therefore used to test the quality of all pharmaceuticals certified by the American Food and Drug Administration (FDA), and similar rules apply in the Asian countries as well. The result is a billion-dollar industry ensuring that all liquids used in medical treatments are pure.

Free Flow of Blood

I ask to hold a bag full of blood. The bag is cool, far from what I think of as being alive. Gravity has divided the bag's contents into three parts: the bottom part has an intense, deep-red color; on top is the yellow fluid; a white membrane of immune cells divides the two floors. The bag can be refrigerated for a whole month and save three human lives.

The front of the bag carries important messages about what is hiding inside: A, rhesus negative, I read. There is also a three-inch-long tube of blood attached to the bag. The blood in the tube is the same as the blood in the bag and is made ready for engineers to easily test whether the blood in the bag will react with the patient's blood and cause harm. That is to say, everyone's blood is not the same.

The surface of blood cells is covered with thousands of different molecules, and plasma is full of antibodies whose job it is to find foreign cells. If an antibody from the patient reacts to the donor's blood cells, things can go very badly indeed.

And things did go very badly in the old days with some of the attempts to perform blood transfusions. Blood clumps when it comes into contact with air, making transfusion difficult. Inside the body, this can lead to serious reactions. The reason why blood cannot flow freely between people is because of what we now know as blood types.

In the early 1900s, Karl Landsteiner used blood from the employees of his laboratory in Vienna to systematically mix blood cells with plasma. If the surface of the blood cells reacted to an antibody in the plasma, the mixture clumped together. Inside a body, clumps of blood can lead to arterial blockage and the release of breakdown product, which can damage the kidneys. Landsteiner's experiments showed that blood from some of his colleagues never clumped (those we now know as type O). The other blood types are distinguished by whether the surface molecule of the blood cells are of type A or B. People with blood type AB have both type A and B on the surface of blood cells.

Together with the ABO classification, the Rh blood group system is the most well-known system for differentiating between blood types, but they are neither the only tests nor the most complex. In 2014, the number was expanded to as many as thirty-three different ways to classify blood. ABO and Rh are the two most relevant because serious complications arise if attention is not paid to them during a blood transfusion.

In 2016, the blood bank at Haukeland Hospital received 21,247 bags of blood: blood that saves lives each and every day. The new blood also has the potential to change the life of the recipient long after receiving the new blood. Several scientists are on the hunt for chemical signals in the blood that can have incredible—and possibly perpetual—consequences for the person receiving an extra bag of blood.

The Power of Young Blood

Research on blood's potential has a long—and long from straightforward—history. It all started with an invasive experiment 150 years ago.

In the mid-1800s, French physiologist Paul Bert sewed two albino rats together with the goal of creating a shared circulatory system between the two animals. Each of the rats first had a piece of skin removed from its side before Bert sewed the two together like Siamese twins. The wound healed on its own, and the rats' capillaries connected with each other in the process. Blood from one flowed into the other, with two hearts pumping the blood between them. The rats lived on, side by side, and the technique was called parabiosis, which means exactly that: "living beside."

Some of the artificial Siamese rats took the lives of their partner, and many pairs died of immune responses. The experiments were neither technically simple nor ethically unproblematic, and in the 1970s, scientists stopped using the method of sewing laboratory animals together.

In recent years, several research groups with labs in California have revived the technique with Siamese rodents. The new ideas are based on experiments from the 1970s, where some results hinted at old rats sewn together with young rats living longer than expected. Could the blood of the younger partner have been the reason?

In 1999, researcher Amy Wagers wanted to study how stem cells moved around in the blood. She sewed two mice together to investigate whether colored stem cells in one mouse could move into the body of its partner. Her results showed that blood stem cells moved out of the bone marrow of the one mouse and contributed to new bone marrow in the other animal. Inspired by this success, Wagers embarked on a new experiment to study the effect of young blood on old mice. She connected together the blood circulation of young and old mice, where the mice's ages corresponded to people aged twenty and sixty-five. The study's conclusions made big headlines: blood from young mice affected stem cells in the skeletal muscles and liver of old mice; old stem cells started to divide again.

Amy Wagers collaborated with heart and brain specialists to detail the effect of young blood. Together they discovered that young blood was able to reverse an enlarged heart muscle in old mice and help increase protection of brain cells. Young blood also led to an increased number of blood vessels in the brains of old mice. The effect was that the blood supply to the brain increased, and cell division in the brain subsequently increased. Everything suggested that young blood had a dramatic effect on old mice.

But Wagers did not want to draw conclusions about the old mice becoming younger because of the young blood. She believed that the young blood instead contributed to repairing old tissue so that the tissue recovered its function. Although Wagers is restrained, other scientists have much bigger expectations about the potential of their discoveries.

The Elixir of Youth: Four Candidates

In 2013, Wagers managed to identify what it was about the young blood that had had such a spectacular effect. Together with researcher Richard Lee, she found that the protein GDF11 was responsible for the effect of young blood on the heart. GDF11 is a growth factor with several different functions in a normal body. Wagers and Lee produced artificial GDF in the lab, injected it into old mice, and noticed thereafter that it had the same effect on the heart as blood from young mice. Mass production of GDF11 as a supplement was poised to become a money-making success.

But other researchers in recent years have had trouble replicating the results of Wagers and Lee. Among other things, there is disagreement about whether the level of GDF11 actually decreases when mice and people get older. If the level does not decrease, it would no longer be reasonable to attempt to increase it in older people. GDF11 is no longer a leading candidate for the elixir of youth.

Irina and Michael Conboy are behind the research on candidate number two for the elixir of youth in blood: the hormone oxytocin. In 2014, the research duo presented results demonstrating that blocking oxytocin in young mice reduced activity of muscle stem cells. They also found that increased amounts of hormone in old mice led to an increase in cell division. The hormone affects the body in different ways, including with falling in love and breast-feeding. It is not clear whether it will be possible to influence the level of oxytocin in old people without producing complex or surprising side effects.

The third and fourth candidates come from Saul Villeda and Tony Wyss-Coray. These colleagues have carried out several comprehensive experiments to show that blood from young mice helps old mice to both learn more quickly and have better memory. One result showed that the brain cells of old mice created a greater number of new connections between each other because of the young plasma. "One possibility is that *youth factors* from young blood can reverse age-related changes in the brain," they write.

In several interviews, Wyss-Coray discusses the technology of "restarting the aging clock." For the experiment, he used plasma from young people and injected it into the blood of old mice, which improved the memory of the old mice. Candidate number three for the elixir of youth is a protein called TIMP2. Wyss-Coray and his colleagues isolated TIMP2 from umbilical cord blood from newborn babies and found that the TIMP2 protein had a large impact on the brains of old mice. The protein found its way into the mice's brains and helped increase activity in the hippocampus, the area of the brain involved in learning and memory.

The mice that received plasma from the umbilical cord did better on learning and memory tests than the mice that received the control treatment. Pure TIMP2 injected right into the blood of the mice also had a positive effect their brains. The director of the National Institutes of Health (NIH) in the United States comments that "TIMP2 holds promise for further research and maybe also for therapeutic development."

Saul Villeda has launched the fourth candidate for the elixir of youth: an enzyme called Tet2. The first discovery was that the level of Tet2 in the brains of mice decreased as they got older. Young mice, in which scientists had turned off Tet2, aged much too early. If scientists instead increased the level of Tet2, the mice performed better on tests for learning and memory. When old mice were sewn together with young mice, scientists saw that the young blood caused the level of Tet2 in the brains of the old mice to increase. Maybe a central effect of the young blood is an increased level of Tet2.

Now scientists want to test whether these discoveries can have an effect on old people. It is most common to use plasma—blood without blood cells—from young donors. Some use plasma from eighteen- to twenty-year-old donors, while others use plasma from the umbilical cord blood that remains after a normal birth.

In September 2014, Alkahest, Wyss-Corday's company, started a clinical study to test whether it is safe to give Alzheimer's patients plasma from young donors. They planned to include eighteen patients who would receive just under half a liter of plasma from twenty-year-olds once a week for a month. In addition to comprehensive blood tests both before and after, the patients were examined through brain scans and mental performance tests. The study is finished, but the researchers have not yet reported their results.

In South Korea, a similar clinical study is taking place to test the effect of young plasma on the aging of middle-aged patients. A study being done in San Francisco includes patients with a Parkinson's-like illness. Both studies, as well as Alkahest's study on patients with Alzheimer's, are financed by investors or insurance companies. The expectations are very high.

Ambrosia's private clinic in Monterey stands out in that patients pay out of pocket—eight thousand dollars per head. If Jesse Karmazin manages to include six hundred volunteers, that will mean almost five million dollars. For the patients that are participating, the stakes are extra high.

Young Donors

Many have believed young blood to be an elixir of youth, and in 1615, Andreas Libavius wrote that blood from young people was a "fountain of life." He imagined that blood could flow from a young, healthy man over to an older man, and that the young, warm blood would strengthen the old and frail.

According to anecdotes, the Hungarian countess Elisabeth Bathory, known as the world's most prolific female serial killer, bathed in the blood of virgins to stay young. The clergy was also not averse to acting on faith in the power of blood. In 1492, Pope Innocent VIII fell into a coma, and unreliable sources report that the treatment attempted consisted of blood from young donors. A doctor poured blood from three ten-year-old boys right into the pope's mouth to save him. Things did not turn out as the doctor had hoped: the pope and the three children died.

Nevertheless, the idea that blood from young people may contain something that can help us live better—and perhaps also longer—is an attractive one.

The head of the blood bank at Haukeland Hospital, Einar Kristoffersen, is critical of the results from the Ambrosia study, where participants pay their own way. People who pay nearly one hundred thousand Norwegian Kroner for experimental treatment will also have high expectations about the effect of young blood. Kristoffersen explains that the experimental setup makes it so that the self-reported results have little value, and that the objective measurements cannot be trusted.

But suppose that it works, and that young blood can be good for old people. Where will the blood come from? Who will the young donors be?

Kristoffersen chuckles at the suggestion that giving blood could become a regular part of confirmation preparation classes, and emphasizes that young blood poses some clear ethical challenges.

Fluids Straight from the Lab

One alternative would be to make artificial blood and add youth factors in the lab, far away from defenseless teenagers. But the complexity and nuance of the living blood of humans are not reproducible, so 100 percent artificial blood produced in the lab is impossible.

Some of blood's main functions can, however, receive synthetic help.

Red blood cells use hemoglobin to bind oxygen, so several companies are working to produce extra hemoglobin as a supplement for people who need to increase the amount of oxygen in their blood. Could it be possible to isolate hemoglobin from old blood or from the blood of cows in slaughterhouses, or to use hemoglobin produced by bacteria? The goal is to increase the transport of oxygen and improve people's health by injecting extra hemoglobin into their blood vessels. There remain major challenges connected to adequate purification, side effects, and chemical modifications before such methods can be of benefit.

Another vital component is platelets, which plug small holes in the blood vessels. Normal platelets have a short lifespan, so one of the artificial alternatives is to recycle expired platelets. The plan is to divide the used platelets into many small pieces, remove viruses and bacteria, and give the resulting smoothie to someone in need. The effect is not nearly as good as with fresh platelets, but it solves some of the global shortage problem.

The recycling of bodily fluids is also the idea behind the company Kybella. In Hollywood clinics, they offer an injection of artificial bile acid to melt away the double chins of celebrities. Bile breaks down fat in the intestine by turning large fat cells into microscopic drops. Investors presumably thought that bile acid could break down fat in places other than the intestine. The result was a shot full of artificial bile acid. Twenty injections in the double chin surround the fat cells with bile acid, and in the following weeks, the blood and lymphatic system transport the drops of fat away. The cost is between twenty thousand and fifty thousand Norwegian Kroner (about two to six thousand dollars). The treatment works as promised, though the effect varies from small to moderate. The injections can also result in nerve damage in the jaw, leading to a crooked smile.

In contrast to the voluntary injections in Hollywood clinics, the majority of children and adults with diabetes have to tolerate a handful of needle pinches every single day. But around 125 enthusiastic test subjects around the world are testing out a new technology that allows them to avoid the tiresome needle. An artificial pancreas could be the solution, where a sensor under the skin measures blood sugar and is connected to a cell phone app. The technology is not yet formally approved, but the testing shows large potential.

Experimental treatment involves significant risk, also for dead people who would like to be brought back to life someday in the future. In *The Atlantic*, Rose Eveleth explains that "those who are going to die today can be cured tomorrow." While they wait, people are having themselves frozen. The main problem is that water in the cells expands when it freezes, causing the cells to explode. Resuscitation will have fatal consequences.

Consequently, the company Alcor has implemented extensive procedures with the dead bodies before freezing them. Since the bodies have been declared dead, the company can employ methods and technology that are not approved for use on living people. First they inject sixteen different medications to protect the cells from the cold. The next step is to remove as much of the blood and tissue fluid as possible, and to replace them with medical antifreeze.

A surgeon opens the sternum to rinse the main arteries and fill them with antifreeze. Slow cooling over a span of two weeks allows the body to finally be stored at minus 328 degrees Fahrenheit. However, having your arteries filled with antifreeze provides no guarantee that you will experience life in the distant future.

In contrast to artificial bile acid in double chins and digital insulin pumps, research on artificial blood has not been a great success. Already since the early 1990s, the head of the blood bank in Bergen has heard praise of what lab-produced blood was going to do. But even though most of the attempts have failed, Kristoffersen has a plan B: in the future, he wants to cultivate blood in the lab. Not from nothing, but from stem cells.

Living Blood

Inspired by the department of cell therapy at the Radium Hospital in Oslo, Kristoffersen and his colleagues are planning a new, high-tech laboratory where they will cultivate blood cells to save seriously ill patients. For unique patients—those who cannot receive blood from blood bank donors—it will be possible to cultivate the blood they need. Kristoffersen points out that it will be very expensive to produce a liter of blood for a single patient, compared to using blood from donors. But this venture could result in a renaissance of lab-produced blood, where *artificial* is replaced with *stem cell–based*.

Several research groups have worked on charting and automating blood production in the lab. Some scientists use stem cells from umbilical cord blood to cultivate blood cells, while others use stem cells in bone marrow as a starting point. The process is extensive, and the cells need daily supervision over the course of a whole month. A few stem cells produce a pretty small volume of finished blood cells, so the cost of one bag of blood for a patient is still very high.

In the spring of 2017, a research team from Bristol came up with a possible solution. They began with stem cells from bone marrow, but instead of forcing the cells to develop into blood, the scientists made the stem cells immortal in the laboratory. The result was everlasting blood stem cells that grow and divide before they later develop into blood. Even in small amounts, stem cells from bone marrow can be important for patients. It is possible that only a few drops of bone marrow are needed to isolate blood stem cells that can continue to grow in the lab and become a treatment in the future. The researchers behind this study of blood stem cells are very optimistic.

Critics, on the other hand, are cautious, especially because of the technology of turning the stem cells in bone marrow into immortal blood cell producers. Scientists used a cancer-causing gene from the HPV virus to make the immortal stem cells, but they do not believe that this will be a problem for the recipient. Extensive testing and safety measures must be put in place before the lab-produced blood cells can become a comprehensive and important tool. The method must mature before clinical trials can begin.

And if scientists are able to create new blood, why would they stop there? We are headed toward a future where it will be possible to create more than just bodily fluids in the lab.

A Video in DNA

Bodily fluids such as saliva, semen, bone marrow, and blood contain living cells. Scientists study the DNA in cells to find disease, identify risk, and aid in selecting proper treatment. Our knowledge of what bodily fluids contain saves lives each and every day, and the latest breakthroughs indicate that bodily fluids can help extend life. But technology does not stop there. Could it be that knowing what is hiding in bodily fluids can also help create new life?

Since 2013, the gene technology CRISPR has revolutionized biomedical research. CRISPR allows scientists to cut, paste, and change the genetic contents of living cells. The technology is based on a defense system that bacteria use to stop a virus attacking, and in recent years it has become the ace up the sleeve of biomedical research.

Bodily fluids are an easily accessible source of living cells that can be changed with CRISPR. Cells in a drop of blood or a gob of spit can receive new superpowers in the lab. CRISPR can help immune cells from blood find cancer cells in a patient, while stem cells from bone marrow can repair a disease-causing gene defect. Cells in saliva can live on in the laboratory as personal research tools, such as in testing new cancer or Alzheimer's drugs.

The limits to what is possible with our cells is both hard to predict and very malleable. Two recent examples demonstrate the unbelievable potential of using CRISPR to change the DNA of different cells.

In the summer of 2017, a group of American researchers presented the results of a rather spectacular set of experiments with storing data in the DNA of a bacterium. Scientists, led by George Church at Harvard University, used CRISPR to paste new pieces of DNA into the bacterium. The new DNA corresponded with the color intensity of pixels in five still images, which together made up a very short video of a man riding a horse. The bacterium grew and divided while scientists waited for an answer to their question: Would the bacteria retain the information about the black-and-white video in their DNA?

With the help of an advanced computer program, scientists put the information in the five still images from the bacteria's DNA back together. The short video of a man riding a horse was still clearly visible, even after the bacteria had divided dozens of times. The experiment showed that it is possible to use CRISPR to store data in living organisms.

Perhaps CRISPR can be used in the same way to store new information in bodily fluids. Would you want to have your mom's meatball recipe, a declaration of love, or a quote from Beyoncé stored in your blood?

Design Baby

A Chinese research group has also pushed boundaries by using CRISPR to change the DNA of a human embryo. Scientists have demonstrated that it is possible to correct gene defects in an embryo so that it can develop without serious disease. The research has laid the groundwork for repairing genetic disease in unborn babies in the future. The technology will undoubtedly be important for families with serious genetic diseases and will prevent the unnecessary suffering of children.

From using gene technology to repair gene defects, it is a short leap to use the same technology to upgrade and improve normal genes. The concept of designer babies, which has neither democratic nor precise associations, is nevertheless a foreshadowing of the future that lies ahead. The technology will make it possible to carry out long-lasting and heritable changes in the genetic qualities of the embryo.

Of course, it will not be possible to plan in detail all physical, mental, and personal attributes with small genetic adjustments in an embryo. Our twenty thousand genes collaborate with each other, and with nature and nurture, in a much too complex manner to predict all the consequences of new changes; it is unlikely we will arrive at that place. This does not mean that the questions we face are simple. As an example, there is much disagreement about where the boundary lies between gene defect and normal variation.

The potential of CRISPR forces the debate on what the limits should be. What is normal? Should people with chromosome defects and genetic diseases be a part of our diversity? What do we do if rich families in the West buy themselves a genetic future that protects them from having to think about diseases of the poor?

Jennifer Doudna, one of the founders of CRISPR, writes about the major ethical challenges with the new technology in the book *A Crack in Creation*. She states very clearly that there is a need for regulating technological advances that can have undesired consequences. Doudna quotes the director of the NIH: "Evolution has been working toward optimizing the human genome for 3.85 billion years. Do we really think that some small groups of human genome tinkerers could do better without all sorts of unintended consequences?"

Contours indicate a field of science where research makes it possible to carry out spectacular experiments with humans—experiments that no one knows the consequences of.

Eve 2.0

I do not think the first lab-born child is many decades away: an embryo that is first sorted out for sex and risk of hereditary diseases, and then gene edited with CRISPR to improve and upgrade a selection of genes linked to intelligence and physical traits.

The baby begins her life in liquid nourishment in the laboratory before growing and developing for nine months in an artificial uterus mounted on a shelf. Here she lies well protected by half a liter of artificial amniotic fluid.

Just born, and crying, Eve 2.0 receives milk from her mother and tears of joy from her father. Days pass, and baby poo is replaced with bed wetting, crocodile tears, and drops of blood before puberty begins. Pimples, periods, and rings of sweat, then sex for the first time.

Her life consists of hard work, champagne parties, disappointments, good times, and everyday life. Some days are full of snot, diarrhea, urine, and an upset stomach of acid; others are full of adrenaline and sweat from an intense workout and hard work. Juices are flowing, and in her mid-thirties, Eve 2.0 meets the person she wants to spend the rest of her life with. They become the parents of two: a set of twins with small genetic adjustments that secure a future without illness or need, and some well-chosen sequences of DNA for a life that perhaps will not be so short.

Eve 2.0 buys daily doses of an elixir of youth, does yoga, and maintains a vegetarian lifestyle. She turns 110, 120, and then 130 years old before passing away peacefully. Her family cries, their tears falling to the earth. In the blood of her children and grand-children, the history of Eve 2.0 and the technological revolution continues to flow.

DEATH—ONE LAST DROP

When we die, the heart stops pumping blood around the body. The cells do not get any more oxygen or nourishment, so energy burning and heat production cease. With each passing hour, the body gets 0.8 degrees colder, until it reaches the same temperature as its surroundings.

In places where blood disappears from the capillaries, the skin becomes pale and lifeless. Gravity pulls the blood downward, coloring the skin with dark spots. Chemical signals in the blood stop, and fibers lock the body into place like a marble statue. Only once muscle cells are broken down does the marble release its grip around the limbs.

Then the body is eaten, first by itself.

In healthy, living cells, the enzymes that divide sugar, fat, and protein into small pieces are kept under strict control. In a dead body, the enzymes do as they wish and eat what is closest: the cells. Molecular cannibalism breaks down the cell membrane, and water from the cells leaks out. Fluid pools in the body's cavities, and as small sacks in the skin. Sometimes the water mixes with dead blood, sometimes it is transparent. Over time, the top layer of skin detaches from the tissue underneath and sheds off in large flakes.

Microorganisms that consume the remains of our food while we are alive consume our bodies when we die. They start in the intestine and chew away. Their waste is gases that are trapped inside the body, a balloon constantly growing in circumference. The rectum collapses, and gases in the fat underneath the skin and in the scrotum have no way out. The pressure builds, and the skin is stretched. Around the mouth and genitals are also numerous bacteria that cause the tongue, lips, and penis to swell.

Fluid from cells mixes with the bacteria and remains of intestine, pancreas, lungs, and fat: an organ soup with lots of activity. Bacteria constantly move on to the next meal. Then come the flies, whose larvae are like small, crawling grains of rice eating and growing inside the dead body.

Unless a dead body is cared for and preserved, or congealed in a cold, dry environment, it eventually collapses and spills outward. Only the skeleton remains, for the time being.

The liquid body is absorbed into the earth and the fluid becomes a part of nature.

Back to the start.

THANKS

I owe a big thanks to everyone who contributed to the contents of this book. A special thanks to the courageous Ingrid Lunde for sharing her story with me. An extra thanks to the researchers at St. Olav's Hospital for letting me donate and document what happened with my bone marrow: Therese Standal, Tobias S. Slørdahl, Lill Anny Gunnes Grøseth, Anne-Marit Sponaas, and Anders Sundan.

Many thanks to everyone who agreed to be interviewed, or contributed in other ways with stories about their own bodily fluids or those of others: Marie E. Rognes, Marius Johansen, Henrik J. Henriksen, Gro Sæther, Kari Husabø (1924–2018), Anders Debes, Håvard Aalmo, Tom Luka, Astrid Aksnes, Ellen Økland Blinkenberg, Einar Kristoffersen, Line Nybakken, Heidi Konestabo, Tone Gadmar, Ole Ivar Burås Storø, Atle Hunnes Isaksen, James Boyda, Loretta Ramos, Ole Kristian Drange, Idun Husabø, Tone Nordbø, Eirik Lehre, Emma Manin, and Angeliki Dimak-Adolfsen.

Thanks to test readers Katharina Vestre, Øyvind Rolland, and Eivind Torgersen for enthusiastic feedback. Thanks to the professional consultants for important corrections and input. If a mistake or two should still remain in the book, it is by the author and none other.

Thanks also to the hard-working and inspirational group in Stallen Freelance Collective.

Finally, a huge thanks to Jon Olav, my toughest critic and biggest fan.

Skål!

Oslo, June 2018.

Åsmund Husabø Eikenes

REFERENCES

SPILLAGE

Aldersey-Williams, Hugh. *Anatomies: A Cultural History of the Human Body*. New York, USA: W.W. Norton & Company Inc., 2013.

Aspelund, A., S. Antila, S. T. Proulx, T. V. Karlsen, S. Karaman, M. Detmar, H. Wiig, and K. Alitalo. 2015. "A dural lymphatic vascular system that drains brain interstitial fluid and macromolecules." *J Exp Med* 212 (7): 991–99. doi: 10.1084/jem.20142290.

Bolt, Nina. *Blod, sved og tårer*. Oslo, Noreg: Tiderne skifter, 1998.

Bucchieri, F., F. Farina, G. Zummo, and F. Cappello. 2015. "Lymphatic vessels of the dura mater: a new discovery?" *J Anat* 227 (5): 702–3. doi: 10.1111/joa.12381.

Cooper, Chris. *Blood: A Very Short Introduction*. Oxford, Storbritannia: Oxford University Press, 2016.

Hill, Lawrence. *Blood: A Biography of the Stuff of Life*. London, England: Oneworld Publication, 2013.

Hippocrates, and Eirik Welo. *Om legekunsten*. Oslo, Norway: De norske bokklubbene, 2000.

Iliff, J.J., M. Wang, Y. Liao, B.A. Plogg, W. Peng, G.A. Gundersen, H. Benveniste, et al. 2012. "A paravascular pathway facilitates CSF flow through the brain parenchyma and the clearance of interstitial solutes, including amyloid beta." *Sci Transl Med* 4 (147):147ra111. doi: 10.1126/scitranslmed.3003748.

Lefrançais, E., G. Ortiz-Muñoz, A. Caudrillier, B. Mallavia, F. Liu, D.M. Sayah, E.E. Thornton, et al. 2017. "The lung is a site of platelet biogenesis and a reservoir for haematopoietic progenitors." *Nature* 544 (7648):105–9. doi: 10.1038/nature21706.

Louveau, A., I. Smirnov, T.J. Keyes, J.D. Eccles, S.J. Rouhani, J.D. Peske, N.C. Derecki, et al. 2015. "Structural and functional features of central nervous system lymphatic vessels." *Nature* 523 (7560):337–41. doi: 10.1038/nature14432.

Mezey, Éva, and Miklós Palkovits. 2015. "Forgotten findings of brain lymphatics." *Nature* 524:415. doi: 10.1038/524415b.

Molteni, Megan. 2017. "Scientists Build a Menstrual Biochip That Does Everything But Bleed." *Wired*. Read June 28, www.wired.com/2017/03/scientists-build-menstrual-biochip -everything-bleed/.

Nedergaard, Maiken. 2013. "Neuroscience. Garbage truck of the brain." *Science* 340 (6140):1529–30. doi: 10.1126/science .1240514.

Nedergaard, Maiken, and Steven A. Goldman. 2016. "Brain Drain." *Scientific American* 314 (3):44–49. doi: 10.1038/scientificamerican 0316-44.

Orešković, Darko, and Marijan Klarica. 2015. "The controversy on choroid plexus function in cerebrospinal fluid production in humans: how long could different views be neglected?" *Croatian Medical Journal* 56 (3):306–10. doi: 10.3325/cmj.2015.56.306.

Porter, Roy. *Blood and Guts: A Short History of Medicine.* London, England: Penguin Books, 2003.

Schutten, Jan Paul. *Den fantastiske kroppen.* Omsett av Bodil Engen. Oslo, Noreg: Spartacus, 2015.

Shwayder, Maya. 2011. "Debunking a myth." *The Harvard Gazette.* Read March 12, 2018. https://news.harvard.edu/gazette /story/2011/04/debunking-a-myth/.

Torgersen, Trond-Viggo. *Kroppen.* Oslo, Noreg: H. Aschehoug & Co. (W. Nygaard), 2003.

Xiao, S., J.R. Coppeta, H.B. Rogers, B.C. Isenberg, J. Zhu, S.A. Olalekan, K.E. McKinnon, et al. 2017. "A microfluidic culture model of the human reproductive tract and 28-day menstrual cycle." *Nat Commun* 8:14584. doi: 10.1038/ncomms14584.

Xie, L., H. Kang, Q. Xu, M.J. Chen, Y. Liao, M. Thiyagarajan, J. O'Donnell, et al. 2013. "Sleep drives metabolite clearance from the adult brain." *Science* 342 (6156):373–77. doi: 10.1126 /science.1241224.

SEX

Baker, Robin. *Sperm Wars*. New York, USA: Basic Books, 1996.

Birkhead, Tim. *Promiscuity: An Evolutionary History of Sperm Competition*. Boston, USA: Harvard University Press, 2000.

Brochmann, Nina and Ellen Støkken Dahl. Gleden med skjeden. Oslo, Norway: H. Aschehoug & Co. (W. Nygaard), 2017.

Chivers, M. L., M. C. Seto, M. L. Lalumière, E. Laan, and T. Grimbos. 2010. "Agreement of self-reported and genital measures of sexual arousal in men and women: a meta-analysis." *Arch Sex Behav* 39 (1):5–56. doi: 10.1007/s10508-009-9556-9.

Chivers, M. L., M. C. Seto, and R. Blanchard. 2007. "Gender and sexual orientation differences in sexual response to sexual activities versus gender of actors in sexual films." *J Pers Soc Psychol* 93 (6):1108–21. doi: 10.1037/0022-3514.93.6.1108.

Deutch, Albert. "Hva Kinsey fant om kvinnen". I Kinsey-rapportene i søkelyset, s. 7–22. Oslo: Dreyer, 1954.

Hippocrates, and Eirik Welo. *Om legekunsten*. Oslo, Norway: De norske bokklubbene, 2000.

Kort, Remco, Martien Caspers, Astrid van de Graaf, Wim van Egmond, Bart Keijser, and Guus Roeselers. 2014. "Shaping the oral microbiota through intimate kissing." *Microbiome* 2 (1):41. doi: 10.1186/2049-2618-2-41.

Partridge, E.A., M.G. Davey, M.A. Hornick, P.E. McGovern, A.Y. Mejaddam, J.D. Vrecenak, C. Mesas-Burgos, et al. 2017. "An extra-uterine system to physiologically support the extreme premature lamb. " *Nat Commun* 8:15112. doi: 10.1038/ ncomms15112.

Roach, Mary. *Bonk: The Curious Coupling of Science and Sex*. New York, USA: W.W. Norton & Company Inc., 2008.

Suarez, S.S., and A.A. Pacey. 2006. "Sperm transport in the female reproductive tract." *Human Reproduction Update* 12 (1):23–37. doi: 10.1093/humupd/dmi047.

Vestre, Katharina. *Det første mysteriet*. Oslo. Norway: H. Aschehoug & Co. (W. Nygaard), 2018.

Vogel, Steven. *Vital Circuits: On Pumps, Pipes, and the Workings of Circulatory Systems*. New York, USA: Oxford University Press, 1992.

Veale, D., S. Miles, S. Bramley, G. Muir, and J. Hodsoll. 2015. "Am I normal? A systematic review and construction of nomograms for flaccid and erect penis length and circumference in up to 15,521 men." *BJU Int* 115 (6):978–86. doi: 10.1111/bju.13010.

Wittkamp, Volker. Ett skritt foran. Oslo, Noreg: Pax forlag AS, 2016.

Zervomanolakis, I., H.W. Ott, D. Hadziomerovic, V. Mattle, B.E. Seeber, I. Virgolini, D. Heute, et al. 2007. "Physiology of upward transport in the human female genital tract." *Ann N Y Acad Sci* 1101:1–20. doi: 10.1196/annals.1389.032.

FOOD

Al-Shehri, S.S., C.L. Knox, H.G. Liley, D.M. Cowley, J.R. Wright, M.G. Henman, A.K. Hewavitharana, et al. 2015. "Breastmilk-Saliva Interactions Boost Innate Immunity by Regulating the Oral Microbiome in Early Infancy." *PLoS One* 10 (9):e0135047. doi: 10.1371/journal.pone.0135047.

Bloudoff-Indelicato, Mollie. 2015. "Ancient Alcoholic Drink's Unusual Starter: Human Spit." *National Geographic*. Read August 14, 2017. https://www.nationalgeographic.com/culture/article/ancient-alcoholic-drinks-unusual-starter-human-spit.

Buser, Genevieve, S. Mató, A.Y. Zhang, B.J. Metcalf, B. Beall, A.R. Thomas. 2016. "Notes from the Field: Late-Onset Infant Group B Streptococcus Infection Associated with Maternal Consumption of Capsules Containing Dehydrated Placenta—Oregon, 2016."

MMWR Morb Mortal Wkly Rep 2017; 66:677–78. doi: http://dx.doi.org/10.15585/mmwr.mm6625a4.

Colen, Cynthia. G., and D.M. Ramey. 2014. "Is breast truly best? Estimating the effects of breastfeeding on long-term child health and wellbeing in the United States using sibling comparisons." *Soc Sci Med* 109:55–65. doi: 10.1016/j.socscimed.2014.01.027.

European Union, Court of Justice. 2017. "Purely plant-based products cannot, in principle, be marketed with designations such as 'milk', 'cream', 'butter', 'cheese' or 'yoghurt', which are reserved by EU law for animal products." Read August 8, 2017. https://curia.europa.eu/jcms/upload/docs/application/pdf/2017-06/cp170063en.pdf.

Hinde, Katie. 2016. "What we don't know about mother's milk." TED. Viewed August 8, 2017. https://www.ted.com/talks/katie_hinde_what_we_don_t_know_about_mother_s_milk.

Holloway, Clint. 2015. "10 Highlights From 'Hannibal' Creator Bryan Fuller and Star Hugh Dancy's Comic-Con Panel." IndieWire. Read August 14, 2017. http://www.indiewire.com/2013/07/10-highlights-from-hannibal-creator-bryan-fuller-and-star-hugh-dancys-comic-con-panel-36508/.

Johnson, Carolyn Y. 2016. "The breastfeeding story is more complicated than you think." Washington Post. Read August 7, 2017. https://www.washingtonpost.com/news/wonk/wp/2016/02/23/what-youve-read-about-breastfeeding-may-not-be-true/.

Lafrance, Adrienne. 2015. "About That Breastfeeding Study." The Atlantic. Read August 7, 2017. https://www.theatlantic.com/health/archive/2015/03/about-that-breastfeeding-study/388309/.

NRK Ekko. 2017. "Abels tårn om morsmelk, influensa og spytt." Listened to August 8, 2017. https://radio.nrk.no/podkast/ekko_-_et_aktuelt_samfunnsprogram/sesong/201702/l_702d00fc-90f4-4c0d-ad00-fc90f43c0d52?utm_source=thirdparty&utm_medium=rss&utm_content=podcast%3Al_702d00fc-90f4-4c0d-ad00-fc90f43c0d52.

Roach, Mary. *Gulp. Adventures on the Alimentary Canal.* New York, USA: W.W. Norton & Company Inc., 2013.

Schutt, Bill. *Eat Me: A Natural and Unnatural History of Cannibalism.* London, Great Britain: Profile Books, 2017.

The Lancet. 2016. "Breastfeeding: achieving the new normal." *The Lancet* 387 (10017):404. doi: 10.1016/S0140-6736(16)00210-5.

Thompson, Helen. 2012. "An Evolutionary Whodunit: How Did Humans Develop Lactose Tolerance?" NPR. Read August 8, 2017. http://www.npr.org/sections/thesalt/2012/12/27/168144785 /an-evolutionary-whodunit-how-did-humans-develop-lactose -tolerance.

Victora, Cesar G., Bernardo Lessa Horta, Christian Loret de Mola, Luciana Quevedo, Ricardo Tavares Pinheiro, Denise P. Gigante, Helen Gonçalves, and Fernando C. Barros. 2015. "Association between breastfeeding and intelligence, educational attainment, and income at 30 years of age: a prospective birth cohort study from Brazil." *The Lancet Global Health* 3 (4):e199–e205. doi: 10.1016/S2214-109X(15)70002-1.

Xu, Duo, Pavlos Pavlidis, Recep Ozgur Taskent, Nikolaos Alachiotis, Colin Flanagan, Michael DeGiorgio, Ran Blekhman, Stefan Ruhl, and Omer Gokcumen. 2017. "Archaic hominin introgression in Africa contributes to functional salivary MUC7 genetic variation." *Molecular Biology and Evolution.* sx206. https://doi.org/10.1093/molbev/msx206.

PISS

Blackstock, Lindsay K.J., Wei Wang, Sai Vemula, Benjamin T. Jaeger, and Xing-Fang Li. 2017. "Sweetened Swimming Pools and Hot Tubs." *Environmental Science & Technology Letters* 4 (4):149–53. doi: 10.1021/acs.estlett.7b00043.

Bourouiba, Lydia, Eline Dehandschoewercker, and John W.M Bush. 2014. "Violent expiratory events: on coughing and sneezing." *Journal of Fluid Mechanics* 745:537–563. doi: 10.1017/jfm.2014.88.

Bourouiba, Lydia. 2016. "A Sneeze." *New England Journal of Medicine* 375 (8):e15. doi: 10.1056/NEJMicm1501197.

Burian, M., and B. Schittek. 2015. "The secrets of dermcidin action." *Int J Med Microbiol* 305 (2):283–86. doi: 10.1016/j.ijmm.2014 .12.012.

Chu, Jennifer. 2016. "Sneezing produces complex fluid cascade, not a simple spray." *MIT News*. Read June 28, http://news.mit.edu/2016/sneezing-fluid-cascade-not-simple-spray-0210.

Driscoll, Emily, and Luke Groskin. 2017. "Breakthrough: Connecting the Drops." Science Friday. Viewed June 28, 2017. https://www .sciencefriday.com/videos/breakthrough-connecting-the-drops/.

Ender, Giulia. *Sjarmen med tarmen*. Omsett av Benedicta Windt-Val. Oslo, Noreg: Cappelen Damm AS, 2015.

Gilet, T., and L. Bourouiba. 2015. "Fluid fragmentation shapes rain-induced foliar disease transmission." *J R Soc Interface* 12 (104):20141092. doi: 10.1098/rsif.2014.1092.

Hamblin, James. *If Our Bodies Could Talk: A Guide to Maintaining and Operating a Human Body*. New York, USA: Doubleday, 2016.

Helander, H.F., and L. Fändriks. 2014. "Surface area of the digestive tract—revisited." *Scand J Gastroenterol* 49 (6):681–89. doi: 10.3109/00365521.2014.898326.

Joung, Y.S., Z. Ge, and C.R. Buie. 2017. "Bioaerosol generation by raindrops on soil." *Nat Commun* 8:14668. doi: 10.1038 /ncomms14668.

Lok, Corie. 2016. "The snot-spattered experiments that show how far sneezes really spread." Nature. Read June 28, 2017. http://www.nature.com/news/the-snot-spattered-experiments-that -show-how-far-sneezes-really-spread-1.19996.

Nøttveit, Andrea Rygg. 2016. "Desse kyrne lever med hol i magen." Framtida.no. Read August 21, 2017. https://framtida.no /2016/03/30/desse-kyrne-lever-med-hol-i-magen.

Palm, W., and C.B. Thompson. 2017. "Nutrient acquisi-tion strategies of mammalian cells." *Nature* 546 (7657):234–42. doi: 10.1038/nature22379.

Scharfman, B.E., A.H. Techet, J.W.M. Bush, and L. Bourouiba. 2016. "Visualization of sneeze ejecta: steps of fluid fragmentation leading to respiratory droplets." *Experiments in Fluids* 57 (2):24. doi: 10.1007/s00348-015-2078-4.

Trimble, Michael. *Why Humans Like to Cry*. Great Britain: Oxford University Press, 2012.

Wang, Y. and L. Bourouiba. 2017. "Drop impact on small surfaces: thickness and velocity profiles of the expanding sheet in the air." *Journal of Fluid Mechanics* 814:510–534. doi: 10.1017/jfm.2017.18.

DNA

Arthur, Rob. 2016. "What's in Your Genes?" Slate. Read September 25, 2017. http://www.slate.com/articles/health_and_science /medical_examiner/2016/01/some_personal_genetic_analysis _is_error_prone_and_dishonest.html.

Boddy, Jessica. 2017. "FDA Approves Marketing Of Consumer Genetic Tests For Some Conditions." NPR. Read September 25, 2017. http://www.npr.org/sections/ health-shots/2017/04/07 /522897473/fda-approves-marketing-of-consumer-genetic -tests-for-some-conditions.

Cale, Cynthia M., Madison E. Earll, Krista E. Latham, and Gay L. Bush. 2016. "Could Secondary DNA Transfer Falsely Place Someone at the Scene of a Crime?" *Journal of Forensic Sciences* 61 (1):196–203. doi: 10.1111/1556-4029.12894.

Eikeseth, Unni. *Norske forskingsbragder*. Vitskaplege oppdagingar gjennom 150 år. Oslo, Norway: Samlaget, 2016.

Forzano, Francesca, Pascal Borry, Anne Cambon-Thomsen, Shirley V. Hodgson, Aad Tibben, Petrus de Vries, Carla van El, and Martina Cornel. 2010. "Italian appeal court: a genetic predisposition to commit murder?" *European Journal of Human Genetics* 18 (5):519–21. doi: 10.1038/ejhg.2010.31.

Green, R.E., J. Krause, A.W. Briggs, T. Maricic, U. Stenzel, M. Kircher, N. Patterson, et al. 2010. "A draft sequence of the Neandertal genome." *Science* 328 (5979):710–22. doi: 10.1126 /science.1188021.

Halvorsen, Kristin, and Ole Johan Borge. 2017. "Skal det lages en biobank med DNA fra nesten alle nordmenn?" Aftenposten. Read September 25, 2017. https://www.aftenposten.no /meninger/debatt/i/7eOvV/Skal-det-lages-en-biobank-med -DNA-fra-nesten-alle-nordmenn—Kristin-Halvorsen-og-Ole -Johan-Borge.

Kokalitcheva, Kia. 2016. "Theranos to Shift Focus on Upcoming 'Minilab' Device It Will Sell to Clinics." Fortune. Read September 25, 2017. http://fortune.com/2016/10/05/theranos -shuts-down-clinics/.

Kokalitcheva, Kia. 2016. "Theranos CEO Elizabeth Holmes Banned From Operating a Lab." Fortune. Read September 25, 2017. http://fortune.com/2016/07/08/theranos-holmes -banned/?iid=leftrail.

Panofsky, Aaron, and Joan Donovan. 2017. "When Genetics Challenges a Racist's Identity: Genetic Ancestry Testing Among White Nationalists." SocArXiv. August 17. osf.io/preprints/ socarxiv/7f9bc.

Rutherford, Adam. *A Brief History of Everyone Who Ever Lived.* London, Great Britain: Weidenfeld & Nicolson, 2016.

Macrakis, Kristie. *Prisoners, Lovers and Spies.* New Haven, USA: Yale University Press, 2014.

Nakhleh, Morad K., Haitham Amal, Raneen Jeries, Yoav Y. Broza, Manal Aboud, Alaa Gharra, Hodaya Ivgi, et al. 2014. "Diagnosis and Classification of 17 Diseases from 1404 Subjects via Pattern Analysis of Exhaled Molecules." *ACS Nano* 11 (1):112–25. doi: 10.1021/acsnano.6b04930.

Regalado, Antonio. 2017. "Baby Genome Sequencing for Sale in China." MIT Technology Review. Read September 25, 2017. https://www.technologyreview.com/s/608086/baby-genome -sequencing-for-sale-in-china/.

Robbins, Rebecca. 2017. "This company wants to analyze your saliva—to try to predict when you'll die." STAT News. Read September 25, 2017. https://www.statnews.com/2017/03/13 /insurance-dna-death-prediction/.

Stockton, Nick. 2016. "Everything You Need to Know About the Theranos Saga So Far." Wired. Read September 25, 2017. https://www.wired.com/2016/05/everything-need-know -theranos-saga-far/.

EVE

Castellano, J.M., K.I. Mosher, R.J. Abbey, A.A. McBride, M.L. James, D. Berdnik, J.C. Shen, et al. 2017. "Human umbilical cord plasma proteins revitalize hippocampal function in aged mice." Nature 544 (7651):488–492. doi: 10.1038/nature22067.

Collins, Francis. 2017. "Aging Research: Plasma Protein Revitalizes the Brain." NIH Director's Blog. Read September 28, 2017. https://directorsblog.nih.gov/2017/04/25 /aging-research-plasma-protein-revitalizes-the-brain/.

Conboy, I.M., M.J. Conboy, A.J. Wagers, E.R. Girma, I.L. Weissman, and T.A. Rando. 2005. "Rejuvenation of aged progenitor cells by exposure to a young systemic environment." Nature 433 (7027):760–64. doi: 10.1038/nature03260.

Defense Advanced Research Projects Agency. 2013. "Pursuit of Scalable, On-Demand Blood for Transfusions Could Yield Novel Means of Therapeutics Delivery." DARPA. Read October 2, 2017. https://www.darpa.mil/news-events/2013-11-12.

Elabd, C., W. Cousin, P. Upadhyayula, R.Y. Chen, M.S. Chooljian, J. Li, S. Kung, K.P. Jiang, and I.M. Conboy. 2014. "Oxytocin is an age-specific circulating hormone that is necessary for muscle maintenance and regeneration." *Nat Commun* 5:4082. doi: 10.1038/ncomms5082.

Eveleth, Rose. 2014. "For $200,000, This Lab Will Swap Your Body's Blood for Antifreeze." The Atlantic. Read September 29, 2017. https://www.theatlantic.com/technology/archive/2014/08 /for-200000-this-lab-will-swap-your-bodys-blood-for-anti-freeze/379074/.

Gontier, Geraldine, Manasi Iyer, Jeremy M. Shea, Gregor Bieri, Elizabeth G. Wheatley, Miguel Ramalho-Santos, and Saul A. Villeda. "Tet2 Rescues Age-Related Regenerative Decline and Enhances Cognitive Function in the Adult Mouse Brain." *Cell Reports* 22 (8):1974–1981. doi: 10.1016/j.celrep.2018.02.001.

Katsimpardi, L., N.K. Litterman, P.A. Schein, C.M. Miller, F.S. Loffredo, G.R. Wojtkiewicz, J.W. Chen, et al. 2014. "Vascular and neurogenic rejuvenation of the aging mouse brain by young systemic factors." *Science* 344 (6184):630–34. doi: 10.1126/science .1251141.

Kresie, Lesley. 2001. "Artificial blood: an update on current red cell and platelet substitutes." *Proceedings (Baylor University Medical Center)* 14 (2):158–61. doi: 10.1080/08998280.2001.11927754.

Korsnes, Malin Kjellstadli. 2016. "Silje (6) har heimelaga buk-spyttkjertel." NRK.no. Read September 29, 2017. https://www .nrk.no/mr/xl/pappa-laga-ny-bukspyttkjertel-til-diabetessjuke -silje-1.13184479.

Lee, Naomi. 2016. "Technology: pharming blood." *The Lancet.* 387 387 (10037):2496. doi: 10.1016/S0140-6736(16)30800-5.

Liang, Puping, Chenhui Ding, Hongwei Sun, Xiaowei Xie, Yanwen Xu, Xiya Zhang, Ying Sun, et al. 2017. "Correction of ß-thal-assemia mutant by base editor in human embryos." *Protein & Cell* 8:811–822. doi: 10.1007/s13238-017-0475-6.

Loffredo, F.S., M.L. Steinhauser, S.M. Jay, J. Gannon, J.R. Pancoast, P. Yalamanchi, M. Sinha, et al. 2013. "Growth differentiation factor 11 is a circulating factor that reverses age-related cardiac hypertrophy." *Cell* *153* (4):828–39. doi: 10.1016/j. cell.2013.04.015.

Lopez, Lindsey Hunter. 2017. "The Injection That Melts a Double Chin." The Atlantic. Read September 29, 2017. https://www.theatlantic.com/health/archive/2017/06 /kybella-the-injection-that-melts-a-double-chin/529893/.

Madrigal, Alexis C. 2014. "The Blood Harvest." The Atlantic. Read October 2, 2017. www.theatlantic.com/technology /archive/2014/02/the-blood-harvest/284078/.

Maxmen, Amy. 2017. "Questionable 'Young Blood' Transfusions Offered in U.S. as Anti-Aging Remedy." MIT Technology Review. Read September 28, 2017. https://www.technologyreview .com/s/603242/questionable-young-blood-transfusions-offered -in-us-as-anti-aging-remedy/.

Ruckh, J.M., J.W. Zhao, J.L. Shadrach, P. van Wijngaarden, T.N. Rao, A.J. Wagers, and R.J. Franklin. 2012. "Rejuvenation of regeneration in the aging central nervous system." *Cell Stem Cell* 10 (1):96–103. doi: 10.1016/j.stem.2011.11.019.

Scudellari, Megan. 2015. "Ageing research: Blood to blood." Nature. Read September 28, 2017. www.nature.com/news /ageing-research-blood-to-blood-1.16762.

Shipman, S.L., J. Nivala, J.D. Macklis, and G.M. Church. 2017. "CRISPR-Cas encoding of a digital movie into the genomes of a population of living bacteria." *Nature* 547 (7663):345–349. doi: 10.1038/nature23017.

Trakarnsanga, K., R.E. Griffiths, M.C. Wilson, A. Blair, T.J. Satchwell, M. Meinders, N. Cogan, et al. 2017. "An immortalized adult human erythroid line facilitates sustainable and scalable generation of functional red cells." *Nat Commun* 8:14750. doi: 10.1038/ncomms14750.

Villeda, S.A., J. Luo, K.I. Mosher, B. Zou, M. Britschgi, G. Bieri, T.M. Stan, et al. 2011. "The ageing systemic milieu negatively regulates neurogenesis and cognitive function." *Nature* 477 (7362):90–94. doi: 10.1038/nature10357.

Villeda, S.A., K.E. Plambeck, J. Middeldorp, J.M. Castellano, K.I. Mosher, J. Luo, L.K. Smith, et al. 2014. "Young blood reverses age-related impairments in cognitive function and synaptic plasticity in mice." *Nat Med* 20 (6):659–63. doi: 10.1038/nm.3569.

Wright, D.E., A.J. Wagers, A.P. Gulati, F.L. Johnson, and I.L. Weissman. 2001. "Physiological migration of hematopoietic stem and progenitor cells." *Science* 294 (5548):1933–36. doi: 10.1126/science.1064081.

Wyss-Coray, Tony. 2015. "How young blood might help reverse aging. Yes, really." TED. Viewed September 28, 2017. https://www.ted.com/talks/tony_wyss_coray_how_young_blood_might_help_reverse_aging_yes_really.

Yong, Ed. 2017. "Blood From Human Umbilical Cords Can Rejuvenate Old Mouse Brains." *The Atlantic*. Read September 28, 2017. https://www.theatlantic.com/science/archive/2017/04/ blood-from-human-umbilical-cords-can-rejuvenate-old-mouse-brains/523509/.

DEATH

Roach, Mary. *Stiff: The Curious Lives of Human Cadavers*. New York, USA: W.W. Norton & Company Inc., 2003.